CW00725746

Birds of the C

by Hilary Aikman and Colin M——

Published by:
Department of Conservation
Wellington Conservancy, P.O. Box 5086
Wellington, New Zealand

The authors:

Hilary Aikman (haikman@doc.govt.nz) and Colin Miskelly (cmiskelly@doc.govt.nz) are both based in Wellington, and work for the Wellington Conservancy of the Department of Conservation, which has had administrative responsibility for conservation work on the Chatham Islands since 1997. Hilary Aikman provides support to threatened bird recovery programmes on the Chatham Islands. Dr Colin Miskelly manages a team of ten staff, most of whom provide support to conservation work on the Chatham Islands.

Hilary Aikman first visited the Chatham Islands as a volunteer with Chatham Island taiko expeditions in 1987, and again in 1992 and in 1995, working with shore plover (Rangatira) and black robin on Rangatira and Mangere. Colin Miskelly also first visited as a taiko expeditioner (in 1978), and then conducted some of his thesis research on Chatham Island snipe on Rangatira in 1983-86, during a more detailed study of the Snares Island snipe. Both authors have had continuous involvement with Chatham Island threatened bird recovery programmes since 1997, and visit the Chatham Islands several times each year as part of their work.

Cover montage designed by Colin Miskelly and Jeremy Rolfe based on photographs of Chatham Island mollymawk, shore plover, Chatham Island snipe (all Colin Miskelly), Pitt Island shag, parea (both John Kendrick/DOC), Chatham Island taiko (Mike Imber), Chatham Island shag (Rod Morris/DOC), Chatham Island oystercatcher (Peter Moore), and Forbes' parakeet (Dave Crouchley/DOC); background image Rangatira coastline (Jeremy Rolfe). Black robin: photographer unknown.

Title page: Brown skua. Photo: Colin Miskelly.

ISBN 0-478-22565-2

Foreword

BY THE HON. CHRIS CARTER,
MINISTER OF CONSERVATION

Chris Carter near Hapupu,
2003. Photo: Daniel King.

The Chatham Islands are a very special part of New Zealand, and I have greatly enjoyed my four visits there over the last decade. Part of the magic of the islands is their unique birdlife, including about 20 species that occur nowhere else on the planet. It was one of these species (the taiko) that first took me to the Chatham Islands, and more recently I have been privileged to witness first-hand some of the remarkable conservation successes that the Chatham Island community and the Department of Conservation are achieving together. This book is a celebration of both the diversity of Chatham Island birdlife, and of the recovery programmes that have successfully brought eight species back from the brink of extinction.

The dramatic rediscoveries of the black robin and Chatham Island taiko, and the subsequent management of these two critically endangered species, are well known both on the Chatham Islands and to conservationists worldwide. Their stories are recounted here, but equally inspirational are the stories of lesser-known species that have responded to innovative and well-researched management. These include encouraging oystercatchers to nest in car-tyre platforms that can be dragged up the beach away from storm surges, and using neoprene screens over burrow entrances of Chatham petrels to discourage competing species from entering and taking over the burrows. These are stories that deserve to be told to a wide audience, and I applaud the authors and the Department of Conservation both for their commitment to the recovery of the threatened birds of the Chatham Islands, and for their efforts in making the results accessible through this fascinating book. I hope that every Chatham Islander will take justifiable pride in their heritage when reading this book, and acknowledge the efforts being made to preserve the special birds of the Chatham Islands and to bring them back to sites where they can be enjoyed by all.

Foreword

BY PHIL SEYMOUR, CHAIR,
CHATHAM ISLANDS CONSERVATION BOARD

Phil Seymour (left) and his father Ron Seymour with a male Antipodean albatross nesting on their land, 2004.
Photo: Graeme Taylor & Colin Miskelly.

Birds have always been a large part of Chatham Islands culture. Early inhabitants wore albatross and parakeet feather ornaments, and harvested several species as food. More recently, bird names have been used to identify the birth-place of Chatham Islanders ("Weka" identifying those born on the Chatham Islands, compared to "Kiwi" for the New Zealand-born), while black robin and taiko appear on souvenirs, local currency and beer brand labels. Conservation work on birds of the Chatham Islands, notably black robin and taiko, has provided much of the international profile of the Chatham Islands. Although less well known, management of other species, including Chatham petrel, parea, and Chatham Islands oystercatcher, has also resulted in dramatic increases in their populations.

Since the establishment of the Department of Conservation in 1987, the Chatham Islands' community has become more involved with, and more informed about, their natural and historic resources. Many Chatham Islands landowners have committed areas of land as reserves, covenants and kawenata, which provide important habitat for threatened birds. Pitt Islanders have been especially fortunate, with the community able to participate in transfers and releases of Chatham petrel and black robin on to predator-fenced private land. Conservation work also provides Chatham Islanders with jobs, as permanent staff or contractors to the Department of Conservation.

Through my involvement with the Chatham Islands Conservation Board I have had, along with other Board members, the opportunity for input into recovery plans to save endangered birds. I have also witnessed the skill, dedication and commitment of both Department of Conservation staff and Chatham Islanders to this cause. I hope that some of the techniques developed could, in future, be used to allow a return to sustainable harvest of some of the less threatened species. This publication highlights the successes of some species recovery programmes, and describes species that require further recovery management. I am very pleased that this book makes this information available to Chatham Islanders and others interested in the birdlife of the islands. Continued understanding and knowledge will ensure that the Chatham Islands community and the Department of Conservation can collaboratively manage the long term survival of our birds for the enjoyment of future generations.

CONTENTS

Acknowledgements

We owe a great deal of thanks to the many Chatham Island and Pitt Island landowners who have made the effort to protect wildlife habitat through formal Conservation Covenants or private initiatives. All but two of the threatened birds described here occur on private land, and some entirely so. We also acknowledge the many fishermen and landowners who have provided logistic support to conservation programmes. Many visitors and residents of the Chatham Islands have provided information about birds and their distributions, particularly Dave Bell, Mike Bell, Adam Bester, Shaun O'Connor, Paul Scofield, and Graeme Taylor. We thank the following for providing photographs: Adam Bester, Bill Carter, Brian Chudleigh, Reg Cotter, Dave Crouchley, Mike Danzenbaker, Helen Gummer, Barry Harcourt, Mike Imber, John Kendrick, Daniel King, the late Fred Kinsky, Mary McEwen, John Mason, Don Merton, Geoff Moon, Peter Moore, Rod Morris, the late Peter Morrison, Alan Munn, Ron Nilsson, Ralph Powlesland, Peter Reese, Christopher Robertson, Jeremy Rolfe, Stella Rowe, Paul Scofield, T. Smith, the late Mike Soper, Richard Suggate, Graeme Taylor, Russell Thomas, Dick Veitch, Geoff Walls, and the late Gavin Woodward. Access to images held by the Department of Conservation was facilitated by Ferne McKenzie. These images have the label "DOC" alongside the photographer's name. Permission to reproduce *rakau momori* images, including the "Unique to Chatham Islands" logo, was obtained from the Hokotehi Moriori Trust; Kate Button facilitated our gaining approval for reproduction of three images held in the Museum of New Zealand Te Papa Tongarewa collection, including image I.006411 (*waka korari*). Approval to reproduce Christina Jefferson's depictions of *rakau momori* images was granted by the editor of the *Journal of the Polynesian Society*. We acknowledge the role that our colleague John Sawyer had in initiating the writing of this book. Alan Tennyson and Trevor Worthy provided advice on the extinct birds of the Chatham Islands, and Brian Bell, Mike Bell, Alison Davis, Jaap Jasperse, Shaun O'Connor, Christopher Robertson, John Sawyer, Paul Scofield and Graeme Taylor provided helpful comments on parts or all of the manuscript. We are indebted to The Hon. Chris Carter, and Phil Seymour for providing Forewords. This book would not have been produced without huge input from Jeremy Rolfe, including map production, image editing, page layout, and liaison with printers.

Scope and format of this book

This book provides full coverage and photographs of all 18 endemic species and subspecies of birds surviving on the Chatham Islands, plus an additional three species that are all-but confined to the Chatham Islands. Less detailed information (and fewer photographs) are provided for the remaining 47 breeding species (including 16 introduced species) plus four non-breeding regular migrants. The approximate size of each species is indicated alongside its heading. The measure represents the length from the tip of the bill to the end of the tail.

Species are grouped into broad habitat types demarked with a different coloured strip on the edge of the page. Appendix 1 provides a full list and status of all bird species known from the Chatham Islands, including extinct species and those species covered in the text sections. If you are interested in identifying vagrant bird species on the Chatham Islands, we recommend *The Field Guide to the Birds of New Zealand* (Heather & Robertson 1996 and subsequent editions).

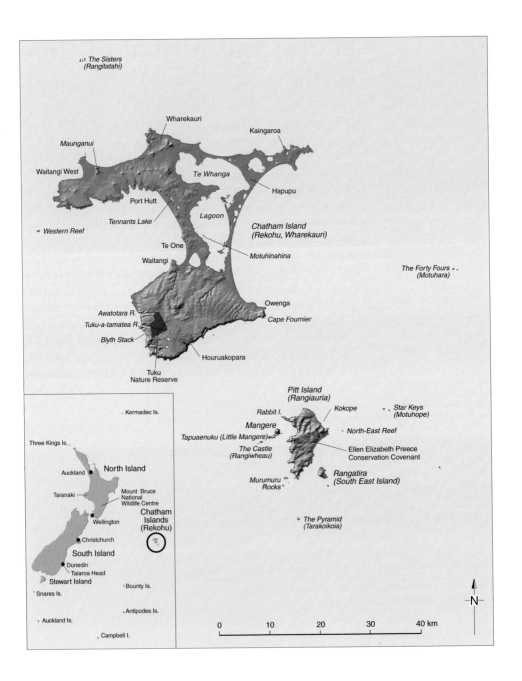

The Sisters
(Rangitatahi)

Wharekauri

Kaingaroa

Maunganui

Waitangi West

Te Whanga

Hapupu

Port Hutt

Lagoon

Tennants Lake

Chatham Island
(Rekohu, Wharekauri)

Western Reef

Te One

Waitangi

Motuhinahina

The Forty Fours
(Motuhara)

Owenga

Awatotara R.

Cape Fournier

Tuku-a-tamatea R.

Blyth Stack

Houruakopara

Tuku
Nature Reserve

Pitt Island
(Rangiauria)

Rabbit I.

Kokope

Star Keys
(Motuhope)

Mangere

Tapuaenuku (Little Mangere)

North-East Reef

The Castle
(Rangiwheau)

Ellen Elizabeth Preece
Conservation Covenant

Murumuru
Rocks

Rangatira
(South East Island)

The Pyramid
(Tarakoikoia)

Kermadec Is.

Three Kings Is.

Auckland

North Island

Taranaki

Mount Bruce
National
Wildlife Centre

Wellington

Chatham
Islands
(Rekohu)

Christchurch

South Island

Dunedin

Taiaroa Head

Stewart Island

Bounty Is.

Snares Is.

Antipodes Is.

Auckland Is.

Campbell I.

N

0   10   20   30   40 km

Introduction

Chatham Islands' birdlife is renowned internationally for two major reasons: the high number of bird species that breed nowhere else on earth, and two remarkable stories of rediscovery and population recovery that serve as beacons of hope and inspiration for conservationists everywhere. This book provides both a summary of the diversity of breeding birds on the Chatham Islands, and information on successes and failures in efforts to ensure the survival of all those species unique to the Chatham Islands.

At least 52 native bird species breed on the Chatham Islands, and 18 of these occur nowhere else (21 if you include two albatross species that have tiny populations elsewhere, and the New Zealand shore plover recently reintroduced to mainland New Zealand). Most Chatham Island bird species evolved from New Zealand forms that colonised the Chatham Islands by flight. It is unlikely that there was ever a land connection to mainland New Zealand, which explains the absence of flightless New Zealand landbirds such as moa, kiwi and kakapo. Although basement rocks on the Chatham Islands are many millions of years old, there is increasing evidence that the islands were submerged until only 1–2 million years ago, and so Chatham Island bird species have colonised and evolved relatively recently. This short history of bird presence and evolution on the Chatham Islands explains why only three species were considered endemic at the genus level (two rails and a duck, all extinct).

Chatham Island rail *Cabalus modestus*, extinct since c.1900. This was the only member of its genus, and was confined to the Chatham Islands. Painting by G.E. Lodge, courtesy of Te Papa, Wellington.

Chatham Island birds have been hard hit by the changes following human colonisation over the last 500 years. About 14 species are believed to have become extinct following Moriori colonisation (Appendix 1), and are now known only from their bones found in middens and natural accumulation sites. Some of these extinct species are yet to be formally described, including a gadfly petrel, a crested penguin, a kaka parrot, and a shelduck. A further seven species became extinct following European and Maori colonisation; four of these occurred nowhere else (a bellbird, a fernbird and two rails). The 21 extinct species represent about 35% of the original bird fauna of the Chatham Islands. Like on other oceanic islands, the main causes of bird extinctions were introduced predators (especially rats and feral cats), their devastation compounded by human harvest and habitat destruction. It is only recently that further extinctions have been prevented by conservation measures, and some species' populations have begun to increase again.

Undoubtedly the most famous Chatham Island bird is the black robin. The story of how this little bird recovered from near extinction (there were only five black robins in existence in 1980) to its current population of over 200 birds is a remarkable tale of persistence, passion, courage and extraordinary good luck. But this has been by no means the only good news in the ongoing battle to save Chatham Island birds from extinction. In 1978, the ornithological world was astounded by the rediscovery of the Chatham Island taiko, a bird unseen by the scientific community for 110 years. It took another ten years before the first taiko burrow was discovered, and ongoing careful management has built up the known population of this critically endangered seabird to about 15 breeding pairs in 2004. While these two icon species have dominated popular understanding of bird conservation work on the Chatham Islands, there are many other examples of species that have responded well to conservation management over the last 30 years.

Rescuing the black robins for transfer to Mangere required scaling the sheer walls of Tapuaenuku. Photo: Don Merton, DOC.

The arrival of the new millennium was rightly celebrated on the Chatham Islands as the first inhabited landmass to greet the morning sun. The new millennium has also signalled a turning point in the recovery of threatened birds on the Chatham Islands. Considerable effort during the 1950s to 1990s was focused on preventing the extinction of about eight endangered bird species, but almost all of this work was undertaken at sites that were inaccessible or off-limits to Chatham Islanders and visitors to the islands. Recent successes and technological advances have started to change this, with Chatham Island oystercatcher and parea (Chatham Island pigeon) now readily seen on parts of main Chatham Island; and efforts since 2002 to restore black robin and Chatham petrel to a predator-fenced site on Pitt Island. With time, adequate resourcing, and support from the local community, viable populations of most of the unique birds of the Chatham Islands should re-establish at secure, accessible sites on the two main islands.

Frano Lanauze, Jack Moffat, Max Gay, Hamish Gregory-Hunt, Yvonne Gregory-Hunt, and Mark Preece release black robins on Pitt Island, September 2002. Photo: Don Merton, DOC.

Moriori, Maori and the birds of the Chatham Islands

Moriori, the first people of the Chatham Islands, are believed to have arrived from New Zealand about 500 years ago. Birds featured prominently in their culture and diet, and the bones of over 50 species have been identified from midden deposits. Species harvested in large numbers included Chatham Island taiko, sooty shearwater, diving petrel, Chatham petrel, parea, Dieffenbach's rail, Hawkins' rail, two species of penguin, tui, parakeets, snipe, black swan and ducks. Methods of capture included extracting nestling petrels from their burrows, spearing parea, other forest birds, and rails, and snaring waterfowl and rails. Following the loss of albatross colonies on Chatham and Pitt Islands, Moriori males proved their manhood during dangerous birding expeditions to the outlying Forty Fours, The Sisters, and The Pyramid. They used *waka korari* (wash-through rafts) as there were no trees suitable for manufacture of dugout canoes. Albatross and mollymawk chicks (*hopo*) were highly valued as food, they featured in ritualistic incantations, their feathers were worn as

3

personal adornment (*raukura*) and attached to *waka korari*, and albatross bones were prized for the manufacture of needles, bird spear tips, and other tools. Parakeet feathers were also worn. Moriori carved images of albatrosses, ducks, and other stylised birds on to limestone caves and living kopi (karaka) trees, indicating their importance to *tchakat henu* (indigenous) culture. The wealth of natural food resources, including birds, present on the Chatham Islands before European contact is indicated by the fact that an estimated 2000 Moriori (three times the current Chatham Islands population) were living without agriculture when the islands were "discovered" by Lieutenant Broughton in 1791.

Much of *tchakat henu* traditional knowledge was lost following the invasion by Taranaki iwi in 1835. However, Ngati Mutunga adopted Moriori traditions and tapu during trips to harvest *hopo*, but with the difference that the more robust European-design boats then available allowed more frequent trips, and larger numbers of chicks to be taken; 1280 Chatham Island mollymawk chicks were taken in one documented visit to The Pyramid. Preserved in their own fat, large quantities of albatross chicks, ducks and muttonbirds were shipped back to Taranaki Maori during the New Zealand Wars (1860–1871) and as late as the Parihaka campaign (1880–1881). *Raukura* came to be used to signify loyalty to Parihaka prophet Te Whiti O Rongomai, and remain a highly valued taonga both on the Chatham Islands and in Taranaki. Large numbers of taiko nestlings were also harvested, including about 1000 from the Tuku valley in 1903.

Following re-introduction of black swans (1890), and introduction of weka (1905), both species became highly valued food sources for Maori, Moriori, and European settlers. Up to 40,000 swan eggs were taken annually when the population reached its peak in the 1950s, and several thousand weka are still legally harvested each year.

Bird images from *rakau momori* (Moriori tree carvings). Based on depictions and photographs by Christina Jefferson; reproduced with permission of the *Journal of the Polynesian Society*.

A history of bird research and conservation on the Chatham Islands

European naturalists were surprisingly slow to visit the Chatham Islands, arriving about 50 years after first European contact. Ernst Dieffenbach visited the main island in 1840, where he collected the only known specimen of Dieffenbach's rail before it became extinct. By the time Henry Travers and Charles Traill visited in 1867–72, many bird species were confined to the smaller outlying islands, as rats and cats were abundant on

The unique specimen of Dieffenbach's rail was collected in 1840; soon after this the species became extinct. Painting by G.E. Lodge, courtesy of Te Papa, Wellington.

SIR CHARLES FLEMING

Charles Fleming (1916–87) first visited Chatham Island as a schoolboy in 1933. He returned with fellow university student Graham Turbott in 1937, and they teamed up with Kaingaroa schoolteacher Allan Wotherspoon to investigate the birds of the Chatham Islands over the 1937/38 summer. Highlights included landing on The Pyramid, a fortnight camped on Rangatira studying shore plover, and a much anticipated landing on Tapuaenuku (Little Mangere Island) on 2 January 1938, where they re-discovered the black robin and Forbes' parakeet.

Charles Fleming at Tuku Farm, Chatham Island, 1938.
Photo courtesy of Mary McEwen.

Fleming remained a staunch advocate for bird conservation and research on the Chathams for the rest of his life, including lobbying for the purchase of Rangatira and Mangere as reserves, and for support for the black robin recovery programme. He was a member of the New Zealand Fauna Protection Advisory Council almost continuously from 1955, often advising the Wildlife Service on Chatham Islands bird recovery programmes. He returned to the Chatham Islands (with Lady Peg Fleming) in 1977 as a guest of the Wildlife Service during the second transfer of black robins from Tapuaenuku to Mangere, and again in 1984 on a "Fauna PAC" visit to Rangatira and Pitt Island. On his last visit, Sir Charles was delighted to see his early research on shore plover on Rangatira being extended by MSc student Alison Davis.

the main island. In the late 1800s, Rangatira and Mangere became magnets for professional bird collectors, who shot many specimens for museums and wealthy collectors in New Zealand and Europe. By 1900, cats had reached Pitt Island and Mangere, and it appeared that several of the sought-after bird species, including Chatham Island rail, Forbes' parakeet, Chatham Island fernbird, black robin, and Chatham Island bellbird, were extinct. However, in 1938, Charles Fleming, Graham Turbott, and Allan Wotherspoon scaled the precipitous cliffs of Tapuaenuku (Little Mangere Island), where locals suggested some of these species may have hung on. Near the summit, they were delighted to discover tiny populations of both black robin and Forbes' parakeet. Fleming and Turbott also landed on The Pyramid, the only known nesting site of the Chatham Island mollymawk. Here they discovered a new form of fulmar prion, which Fleming described in his classic 1939 publication *Birds of the Chatham Islands*. The Chatham Island mollymawks that they photographed there had previously been named based on specimens collected on The Pyramid by Rollo Beck of the Whitney South Seas Expedition in 1926.

The cats on Mangere exterminated the introduced rabbits there, plus at least 12 native bird species, before they were themselves shot out by visiting sheep-shearers. By the time the New Zealand Wildlife Service focussed their attention on the Chatham Islands in the 1950s and 1960s, both Rangatira and Mangere were farmed, but miraculously were free of rats, cats, possums or other introduced predators. Both islands were purchased by the Crown (with assistance from the Royal Forest & Bird Protection Society for Mangere), farm stock was removed in 1961 (Rangatira) and 1968 (Mangere), and both islands were gazetted as Nature Reserves, with access by permit only. This signalled the beginning of concerted conservation action to safeguard the remaining birds of the Chatham Islands. At this time, Chatham petrel, shore plover, and Chatham Island snipe were all believed to be confined to Rangatira, and Forbes' parakeet and black robin were confined to Tapuaenuku. Chatham Island oystercatcher and parea were also gravely threatened.

An early success story was the reintroduction of Chatham Island snipe to Mangere in 1970, but shore plover transferred at the same time rapidly flew back to Rangatira. A research programme on the black robin found fewer than 20 birds surviving on the top of Tapuaenuku, with both the population and habitat in decline. The news was little better for Forbes' parakeet. Although they had successfully recolonised

DON MERTON

Don Merton first travelled to the Chatham Islands (as a senior fauna conservation officer for the New Zealand Wildlife Service) in 1968. He was closely involved with the research programme on black robins on Tapuaenuku over 1972–76, then became the key figure during the intensive management of the robins in 1976–89, when black robins were transferred first to Mangere then to Rangatira, and their numbers increased from five to more than 80 birds. During this time, Merton developed and modified many techniques not previously used on small songbirds, including cross-fostering, and safe methods to move eggs, chicks and adults between islands on local fishing boats, often in rough sea conditions. Merton shifted focus to the Department of Conservation's kakapo recovery programme during the 1990s, but returned to the Chatham Islands to experience the dawn of the new millennium from the summit of Rangatira. He also participated in the transfers of black robins from Rangatira to Pitt Island in 2002 and 2004.

Don Merton inspecting a nest box containing black robin eggs, with a female Chatham Island tomtit cross-foster parent in attendance, Rangatira 1987. Photo: Don Merton, DOC.

adjacent Mangere, they were hybridising extensively with red-crowned parakeets in the severely modified habitat there. Both black robin and Forbes' parakeet were clearly on the brink of extinction. In response, the Wildlife Service began a long-term planting programme on Mangere to restore forest to the former pasture. Periodic culling was undertaken to remove hybrid and red-crowned parakeets to safeguard the Forbes' parakeets, and in 1976/77 the bold decision was made to transfer the last seven surviving black robins from Tapuaenuku to a tiny forest remnant on Mangere. The development of new cross-fostering techniques using Chatham Island tomtits on Rangatira started a dramatic recovery over the years 1983–89. When the cross-fostering was stopped in 1989, about 80 robins occupied all suitable habitat on both Rangatira and Mangere.

The frequent visits to Rangatira and Mangere by Wildlife Service and then Department of Conservation staff provided opportunities for several university research projects during the 1970s to 1990s. Species investigated included parakeets, brown skua, Chatham Island warbler, snipe, shore plover, and Chatham petrel. Research was also undertaken on Chatham Island oystercatchers on the northern beaches of main Chatham Island, which has assisted design of the current, very successful, predator control programme protecting oystercatcher nests there.

On Rangatira, the critically endangered Chatham petrel was found to be declining through competition by the similar-sized, but vastly more abundant broad-billed prion. Intensive research and innovation led to the development of neoprene (wetsuit material) screens placed over burrow entrances, which the prions are reluctant to enter, but the petrels push through. As a result, over 100 Chatham petrel chicks now fledge each year.

While the Wildlife Service focussed their attention on the birds on the outlying islands, David Crockett began his quest to unravel the mystery of the taiko. Older locals talked about a seabird that formerly bred in burrows under the forest in the south of main Chatham Island, but specimens had never been seen by scientists. Following tantalising glimpses of birds in flight at night in 1973, Crockett and his team succeeded in catching two taiko in 1978. These proved to be Magenta petrels, known from a single specimen shot in the middle of the South Pacific Ocean in 1867. However, the first taiko nest burrows weren't found until 1987/88, after birds were tracked using radio transmitters. The Department of Conservation was formed in 1987, and continues to work with Crockett's Taiko Expeditions, and landowners, to locate and protect taiko breeding burrows. The extensive predator control undertaken to protect taiko has also benefited parea, which are now commonly seen in the Tuku Nature Reserve and adjacent covenants on private land.

DAVID CROCKETT

David Crockett joined the Ornithological Society of New Zealand as a schoolboy in 1950. Through Ron Scarlett at the Canterbury Museum, and Charles Fleming, Crockett developed an interest in the mysterious taiko of the Chatham Islands, and corresponded with former Tuku landowner H.G. Blyth in 1952. He first visited the Chatham Islands in 1969, and started dedicated searches for the taiko in 1972, culminating in the capture of two birds on the night of 1 January 1978. Crockett's "Taiko Expeditions" maintain Taiko Camp on land owned by the Tuanui family; it has been used as a base for 25 expeditions since 1972. Crockett and his teams of OSNZ volunteers and other taiko expeditioners continue to work closely with the Department of Conservation during telemetry programmes every second year, when taiko are radio-tracked in order to locate further breeding burrows. Crockett lives in Whangarei, and was a

David Crockett with the first two taiko caught, 2 January 1978.
Photo: Russell Thomas.

science educator before retiring in 1995. He is a foundation member of both the Chatham Island Taiko Trust and the Taiko Recovery Group, and has become a well-known identity on Chatham Island during his 51 visits over 35 years.

During the 1980s and 1990s, Chatham Island and Pitt Island landowners began to take a more active role in protecting the dwindling forest cover and birdlife of the islands. Since 1983, private landowners have set aside 33 parcels of land as Conservation Covenants or Kawenata. Most of these sites are fenced to exclude domestic and feral stock, and possums are controlled in many by the Department of Conservation and landowners. Collectively, these covenants on private land protect nearly 2500 ha of Chatham Island, and 188 ha of Pitt Island; they hold breeding populations of nine of the endemic bird species of the Chatham Islands.

The construction of a cat and weka-proof fence around part of the Ellen Elizabeth Preece Conservation Covenant on Pitt Island signalled a new approach to managing predator threats to Chatham Island birds. Up to 55 Chatham petrel chicks were transferred to this site from Rangatira each year 2002–04, and they should start returning in 2004/05. The first transfer of 14 black robins took place in 2002. It failed when the birds dispersed beyond the fence, but further efforts are underway to restore this and other threatened bird species to sites where they can be seen by Chatham Islanders and their guests.

The future

Current conservation effort is focused on securing existing populations, and returning some locally extinct species to a predator-fenced site on Pitt Island. This latter concept is likely to be extended to Chatham Island if resources are secured for building predator-proof fences around forest remnants. The Chatham Island Taiko Trust already has funds to build a predator-proof fence to establish a secure site for Chatham Island taiko and Chatham petrel on Chatham Island. This 1–2 ha exclosure will not be large enough to allow re-introduction of forest birds, but a larger fence proposed for northern Chatham Island would potentially allow the re-introduction of Forbes' parakeet, snipe, black robin, tomtit, warbler, tui, and a variety of burrow-nesting seabirds, as well as range expansion of parea and red-crowned parakeet. More intensive and extensive predator and possum control in the forests of the Tuku Nature Reserve and adjacent Conservation Covenants would potentially allow re-introduction of forest birds as well. This site is currently considered too large to enclose in a predator-proof fence, and so ways would need to be found to keep rats, feral cats, weka, possums, pigs, feral cattle and feral sheep at sufficiently low densities for vulnerable birds to be re-established.

Another possibility is to re-introduce species that have become extinct on the Chatham Islands but survive elsewhere in New Zealand: potential candidates include brown teal, shoveler and scaup. A further extension of this idea would be to attempt introduction of species closely related to extinct

MANUEL & EVELYN TUANUI

Left: Manuel Tuanui at Taiko Camp, 1982. Photo: Stella Rowe.
Right: Evelyn Tuanui holding a taiko, Taiko Camp, 1982.
Photo: Reg Cotter.

Manuel Tuanui worked as a farmhand for H.G. Blyth before purchasing his Tuku farm in 1948. Manuel, Evelyn and their family were enthusiastic supporters of the attempts by David Crockett and his team to rediscover the taiko, allowing search teams to camp on their land, and providing much logistic support and encouragement. They were delighted by the capture of the first two taiko in 1978. In 1983, they generously donated the 1238-ha Tuku Nature Reserve to the people of New Zealand, when it became apparent that this was the key site for taiko. About 80 percent of known taiko breeding burrows are located within the Tuku Nature Reserve.

endemic Chatham Island birds, for example Snares crested penguin, paradise shelduck, kaka, Snares Island fernbird, and bellbird. Chatham Island birds already have a role to play in restoring species extinct on the New Zealand mainland, as is happening with buff weka and shore plover. This concept could be extended to one further species if Chatham Island snipe were used to replace the extinct North Island snipe.

The restoration of seabirds to Chatham, Pitt and Mangere Islands, by protecting remnant and colonising populations, and using attraction techniques such as broadcasting calls, and decoys, is likely to be another focus. This is already underway for Chatham Island taiko, Chatham petrel and Antipodean albatross, but could be extended to sooty shearwater, storm petrels, diving petrels, prions, and other albatross species. The necessary predator-fence designs and attraction or re-introduction techniques already exist. All that is needed is agreement on suitable sites, and the funds to build the fences and/or control pest species.

The continuing survival and recovery of the birds of the Chatham Islands is dependent on ongoing vigilance and care to prevent the further spread of predators to the Chatham Islands, and on to outlying

Bruce and Liz Tuanui, Tuku Farm, 1999. Photo: Bill Carter.
Liz Tuanui with Chatham Island akeake seedlings destined
for the Mangere revegetation programme, Tuku Farm, 1999.
Photo: Bill Carter.

BRUCE & LIZ TUANUI

Bruce and Liz Tuanui purchased Tuku Farm following the death of Manuel Tuanui (Bruce's father), and have continued to take an active role in protecting the natural values on their farm and elsewhere on the Chatham Islands. They have established six Conservation Covenants on 153 ha of their and Tuanui family land, and are foundation members of the Chatham Island Taiko Trust, which has raised funds to build a predator-proof fence for taiko within their Sweetwater Conservation Covenant. Liz Tuanui (née Gregory-Hunt) was a Chatham Islands Conservation Board member during 1993–99, and the Board chair for the last two years of her term. She manages a native plant nursery at their home, growing seedlings for the Mangere revegetation programme; the planting on Mangere has been done by the Tuanui family under contract to the Department of Conservation since 1993. Liz has also been a member of the Taiko Recovery Group since 1993.

islands. Of particular concern is the very real risk of farm goats escaping into the forests of southern Chatham Island. Fortunately rabbits never established on the Chatham Islands (apart from briefly on Mangere), and as a result stoats, weasels and ferrets were not introduced; these agile predators are a major cause of population declines of threatened birds on mainland New Zealand. The absence of rats and possums on Pitt Island is the main reason that tui and tomtits have survived there but not on Chatham Island. Rangatira, Mangere and all other outlying islands are free of all these introduced species, as well as cats, mice, weka, pigs and feral stock. The absence of any introduced mammals on these islands is the single reason that ensured the survival of at least 20 bird species that would otherwise be extinct in the Chatham Islands.

Sites to visit

Many endemic birds of the Chatham Islands can be seen around the coastline or in reserves in public ownership. In fact, just walking or driving around the Chathams you are likely to encounter a number of them. However, some endemic birds are so rare that they are found in only one or two places. Others are confined to remote privately owned islands, or nature reserves administered by the Department of Conservation where public access is restricted. Information about access to sites mentioned at the start of each habitat section is available from the Department's Chatham Island Area Office. A number of endemic birds may be found on private land, and permission must be obtained from landowners before access to these areas.

Most endemic and native birds identified in this guide are protected species and should not be handled or disturbed. The Department of Conservation welcomes observations of rare species beyond their known range (as depicted in the distribution maps). Please contact the Chatham Island Area Office in Te One (phone 3050-098) with information about sightings.

Selected bibliography

Many of the primary research papers on Chatham Island birds can be found in *Notornis*, the quarterly journal of the Ornithological Society of New Zealand. Most of these are referenced in the following texts that we have drawn on freely in compiling this book:

A complete guide to Antarctic wildlife: the birds and mammals of the Antarctic continent and Southern Ocean. Hadoram Shirihai, 2002. Alula Press, Knapantie, Finland.

A working list of breeding bird species of the New Zealand region at first human contact. Richard Holdaway, Trevor Worthy & Alan Tennyson, 2001. *New Zealand Journal of Zoology 28*: 119–187.

Action plan for seabird conservation in New Zealand. Parts A & B. Graeme Taylor, 2000. Department of Conservation, Wellington.

Albatross biology and conservation. Graeme Robertson & Rosemary Gales (editors), 1998. Beatty & Sons, Chipping Norton.

Birds of the Chatham Islands. Charles Fleming, 1939. *Emu* 38: 380-413, 492-509.

Chatham Islands ornithology. Richard Holdaway (editor), 1994. Supplement to *Notornis* Vol. 41.

Chatham Islands threatened birds recovery and management plans. Hilary Aikman, Alison Davis, Colin Miskelly, Shaun O'Connor & Graeme Taylor, 2001. Threatened species recovery plans 36-46. Department of Conservation, Wellington.

Classifying species according to threat of extinction. A system for New Zealand. Janice Molloy, Ben Bell, Mick Clout, Peter de Lange, George Gibbs, David Given, David Norton, Neville Smith & Theo Stephens, 2002. Threatened species occasional publication 22. Department of Conservation, Wellington.

Handbook of Australian, New Zealand and Antarctic birds, Vols 1-6. Various editors, 1990-2002. Oxford University Press, Melbourne.

Moriori: a people rediscovered. Michael King, 1989. Viking, Auckland.

New Zealand birds. 2nd edition. W.R.B. Oliver, 1955. Reed, Wellington.

New Zealand threat classification system lists. Rod Hitchmough (compiler), 2002. Threatened species occasional publication 23. Department of Conservation, Wellington.

Reader's Digest Complete book of New Zealand birds. Christopher Robertson (editor), 1985. Reader's Digest, Surry Hills, New South Wales.

The black robin: saving the world's most endangered bird. David Butler & Don Merton, 1992. Oxford University Press, Auckland.

The Chatham Islands: heritage and conservation. Chatham Islands Conservation Board, and Department of Conservation, 1996. Canterbury University Press, Christchurch.

The field guide to the birds of New Zealand. Barrie Heather & Hugh Robertson, 1996. Viking, Auckland.

The history of the Chatham Islands' bird fauna of the last 7000 years - a chronicle of change and extinction. Phil Millener, 1999. *Smithsonian Contributions to Paleobiology 89*: 85-109.

Wader studies in New Zealand. Hugh Robertson (editor), 1999. *Notornis* Vol. 46 (1): 242 p.

Oceanic birds

Sooty shearwaters.
Photo: Colin Miskelly.

Much of the diversity of Chatham Islands birds is in the wealth of albatross and petrel species that spend most of their lives at sea, returning to land only to breed. These oceanic species are all extremely vulnerable to introduced predators, and, on the Chatham Islands, are now almost entirely confined to the predator-free outer islands for breeding. Two of these islands (Rangatira and Mangere) are Nature Reserves with restricted access, and other islands (including The Sisters, The Forty Fours, The Pyramid, and Tapuaenuku) are privately owned. Most species can be seen from a boat in the vicinity of the breeding islands, but note that two of the rarest species (Chatham Island taiko and Chatham petrel) are unlikely to be seen near land until just before nightfall. A handful of species (e.g. Pacific mollymawk, northern giant petrel, sooty shearwater, and sometimes northern royal albatross) are readily seen from coastal headlands.

Northern royal albatross
Diomedea sanfordi

115 cm

NEAR ENDEMIC TO CHATHAM ISLANDS,
NATIONALLY VULNERABLE

Other name: toroa

Breeding
distribution

Right:
Northern royal albatross.
Photo: Christopher
Robertson (DOC).

Northern royal albatross pair
at nest, Little Sister Island.
Photo: Rod Morris (DOC).

Identification

The largest albatross regularly seen at the Chatham Islands, with a wing span of up to 3 metres. White on the head, body and under-wing, and black on the upper surface of the wing. Juveniles have black mottling on the back between the wings. The bill is light pink with black on the cutting edge of the upper mandible.

Distribution and ecology

Almost all northern royal albatrosses (99.5%) breed on the Chatham Islands—on The Forty Fours, Big Sister Island and Little Sister Island, where there is an estimated breeding population of around 6500 pairs. A small number of birds (20–30 pairs) breed at Taiaroa Head on the Otago Peninsula, and a few breed with southern royal albatrosses (*Diomedea epomophora*) on Enderby Island in the Auckland Islands. Successfully breeding northern royal albatrosses lay a single egg every 2 years, as incubation and chick-rearing takes about 11 months. If a breeding attempt fails at the egg or early chick stage, they will re-nest the following season. Royal albatross pair for life and reaffirm their bond with elaborate courtship rituals when they reunite in September–October, at the beginning of each breeding season. Nests are circular mounds of vegetation, small stones or peat.

Northern royal albatrosses forage in the South Pacific Ocean close to New Zealand during the breeding season and, when not breeding, move widely over the Southern Ocean in a circumpolar migration, moving from west to east with the prevailing winds between 30°S and 50°S.

Threats and conservation

The major threat at present appears to be habitat degradation due to severe storm events. A storm in 1985 removed large amounts of soil and vegetation from The Sisters and The Forty Fours, impacting on the nesting material available. As a result, nests were built from stones, or eggs were laid on bare rock, resulting in low hatching success. Habitat degradation was exacerbated by the normally biennial breeding pattern being disrupted by low breeding success, resulting in most of the total breeding population attempting to nest annually. This prevented the habitat from recovering as the high density of birds stripped the remaining vegetation in an attempt to create nests.

To date, few northern royal albatrosses have been captured on tuna long-lines, and there are no records of by-catch from trawl fisheries. The high survival rate of adults and fledglings indicate that this form of mortality is not a major threat. However, because of the large area

Northern royal albatross colony, Big Sister Island. Photo: Christopher Robertson (DOC).

over which the birds forage, they could be at risk from a wide range of pollutants or oil spills. Northern royal albatross have been harvested in large numbers in the past. The illegal harvest of chicks still poses a threat, and there have been a number of incidences of birds being taken illegally over the past 20 years.

The small Taiaroa Head population of northern royal albatross is very accessible, and the breeding biology and population dynamics of the species have been studied closely there since the colony's establishment in 1919. The populations on the Chatham Islands were studied in the 1970s and 1990s. Most of the research was conducted on Little Sister Island and involved an assessment of population dynamics, breeding success, breeding biology and the effects of habitat change on the breeding population. Aerial photographic surveys have been undertaken three times annually to count numbers of breeding pairs, and to determine breeding success at both The Sisters and The Forty Fours colonies. Satellite tracking has been carried out on adults from both Taiaroa Head and The Sisters colonies to monitor movements during the breeding season and dispersal after breeding.

Northern royal albatross colony, Little Sister Island.
Photo: Christopher Robertson (DOC).

Antipodean albatross
Diomedea [exulans] antipodensis   100 cm

NEW ZEALAND ENDEMIC, RANGE RESTRICTED

A form of wandering albatross, the Antipodean albatross is similar in size to the royal albatross, but with more dark feathering on the body. Females are darker than males, being dark brown over most of the body, but with a white face, under-wings and belly. One pair attempted to breed (but failed) on farmland on southern Chatham Island in both 2003 and 2004, and another pair was raising a chick on Pitt Island when this book went to press (June 2004). The bulk of the population breeds on the Antipodes Island 650 km south-southwest of the Chatham Islands.

Breeding distribution

Adult female Antipodean albatross incubating, Chatham Island, April 2003. Photo: Colin Miskelly.

18

White-capped mollymawk
Thalassarche [cauta] steadi 90 cm

NEW ZEALAND ENDEMIC, RANGE RESTRICTED

Breeding
distribution

White-capped mollymawk
incubating, The Forty Fours,
December 1996. Photo:
Christopher Robertson.

Also called shy mollymawk, or shy or white-capped albatross (previously *Diomedea cauta steadi*), the white-capped mollymawk has a white head, neck, body and rump, with black upper-wings, back and tail. The white under-wings have a very narrow black border with a small triangular black notch at the base of the leading edge. The bill is greenish horn with a yellowish tip. Juveniles have dark bills, and grey shading on the head. One pair was recorded incubating on The Forty Fours in December 1991 and 1996, and with a chick in January 1997. Most of the population breeds on the Auckland Islands.

Salvin's mollymawk *Thalassarche salvini* 90 cm

NEW ZEALAND NATIVE, RANGE RESTRICTED

Also called Salvin's albatross (previously *Diomedea cauta salvini*). Salvin's mollymawk is similar in appearance to the white-capped mollymawk, but with pale grey on the crown, face and neck, and a greyer bill with a dark tip to the lower mandible. The under-wing also has more black at the tip. The main population of Salvin's mollymawk breeds on the Bounty Islands, and a few individuals attempt to breed on The Pyramid, where they pair with Chatham Island mollymawks.

Breeding
distribution

Salvin's mollymawk,
The Pyramid.
Photo: Christopher
Robertson (DOC).

19

Chatham Island mollymawk
Thalassarche eremita 90 cm

CHATHAM ISLANDS ENDEMIC, SERIOUS DECLINE

Other name: Chatham albatross

Identification

The Chatham Island mollymawk is now regarded as a full species; formerly it was regarded as a subspecies of shy albatross (*T. cauta*). It is a medium-sized, dark mollymawk (wingspan 2.2 metres) with a dark grey head and back, and white breast, belly and rump. The upper surface of the wings and the tail are a darker grey, and the wings are white underneath, with narrow black borders. The bill is bright orange-yellow with a dark spot at the tip of the lower mandible. Juveniles have more extensive grey plumage, and a grey bill with black tips on both mandibles.

Breeding
distribution

Distribution and ecology

The Pyramid (south of Pitt Island) is the only confirmed breeding site of the Chatham Island mollymawk. A few birds have been seen ashore on the Snares Islands (one egg recorded, which failed to hatch) and at Albatross Island off Tasmania. There are estimated to be about 4500 breeding pairs on The Pyramid each year, but the low survival rate of adult Chatham Island mollymawks indicates that the total population is in decline.

Chatham Island mollymawks are rarely seen near the coasts of

Chatham Island mollymawk with chick,
The Pyramid, December 2001.
Photo: Paul Scofield.

Chatham Island mollymawk colony, The Pyramid, December 2001.
Photo: Paul Scofield

Chatham and Pitt Islands, apparently foraging to the east, south and west of the islands. They spend the winter off the coasts of Chile and Peru, returning to New Zealand waters for the breeding season. Little is known regarding the movement of young birds, between fledging and returning to the breeding colony as adults, but they may remain off the west coast of South America. Chatham Island mollymawks breed annually from September to April; like all albatrosses and petrels they lay a single egg. Nests are a shallow cup on a small pedestal of soil, rock chips, guano and vegetation. In dry, sheltered sites, nest pedestals can last for many years and reach a height of up to 1.5 metres.

Threats and conservation

Chatham Island mollymawks have been caught as by-catch in a wide variety of fisheries, both in New Zealand waters and off the coasts of Chile and Peru. Chicks have been harvested in large numbers in the past. Harvest is now illegal; however, there is anecdotal evidence that small numbers of chicks continue to be taken sporadically. The loss of soil and drying of the island as a result of adverse weather can cause egg mortality for Chatham Island mollymawks, as has been observed with northern royal albatross. Some nest sites near the shore have been lost to encroaching fur seals.

Other than legal protection of the species, no specific conservation actions have been taken for Chatham Island mollymawk. During the 1970s, work was undertaken to estimate the size of the population and aspects of the species' breeding biology. Further research has been conducted during the 1990s to 2003 to determine productivity estimates, age at first breeding, and more accurate population estimates. Satellite transmitters have been used to monitor birds' movements at sea.

Chatham Island mollymawk chick close to fledging, The Pyramid, February 1993. Photo: Graeme Taylor.

Indian yellow-nosed mollymawk
Thalassarche [*chlororhynchos*] *carteri*   75 cm

NEW ZEALAND NATIVE, COLONISER

A small, white-headed albatross with a long, slender black bill with a yellow stripe along the top only, and a reddish tip. A single pair that has bred on The Pyramid in recent years is the only breeding record from the New Zealand region. All other breeding sites are in the southern Indian Ocean.

Breeding distribution

Indian yellow-nosed mollymawk incubating, The Pyramid, December 2001. Photo: Paul Scofield.

22

Pacific mollymawk
Thalassarche undescribed sp.   80 cm

Breeding
distribution

Pacific mollymawk off
Chatham Islands.
Photo: Mike Danzenbaker.

NEAR ENDEMIC TO CHATHAM ISLANDS,
RANGE RESTRICTED

Other names: Pacific albatross, northern Buller's mollymawk

Identification

Until recently known as the northern Buller's mollymawk (*Diomedea bulleri platei*), now regarded as a separate species. It remains formally undescribed because the type specimen for "*platei*" is now known to be a juvenile (southern) Buller's mollymawk. Pacific mollymawks are small albatrosses (2 metre wingspan) with dark grey on the top surface of the wings, back and tail, and a black band around the edge of the under-wing. The head is medium-grey with a silvery-grey crown, and the rest of the body is white. Adults have a black bill with yellow bands along the top and bottom, while the bill is all dark on juveniles.

Pacific mollymawk with
chick, Little Sister Island,
February 1995.
Photo: Christopher
Robertson (DOC).

Distribution and ecology

This is the albatross most frequently seen around the coast and following fishing boats at the Chatham Islands. Pacific mollymawks breed on The Forty Fours, Big Sister Island and Little Sister Island in the Chatham Islands. There is also a small population, of about 20 pairs, on Rosemary Rock in the Three Kings Islands off North Cape. There have been no counts of the largest breeding colonies on The Forty Fours and Big Sister Island. However, on the basis of area of

occupancy, the population is estimated to be 16,000 breeding pairs on The Forty Fours, 1500 pairs on Big Sister Island, and 630–670 pairs on Little Sister Island.

Pacific mollymawks breed annually October–June. The single eggs are laid in pedestal nests on cliff ledges and on steep faces at the top of cliffs on The Sisters, and on the plateau basins on The Forty Fours.

Pacific mollymawk colony, Little Sister Island, February 1995. Photo: Christopher Robertson (DOC).

Threats and conservation

Only one Pacific mollymawk is known to have been caught on a tuna long-line. However, until recently, there has been no observer coverage on long-liners operating on the Chatham Rise, an area used by Pacific mollymawks. Pacific mollymawks regularly follow crayfish boats at the Chatham Islands and take discards from the pots. One has been confirmed killed by trawling. There have been no recorded instances of illegal harvest of Pacific mollymawk chicks. The mollymawks are somewhat less affected by habitat changes than royal albatrosses on the same islands, as they mainly nest on steep cliff margins (The Sisters) or rough plateau basins (The Forty Fours) where there is more soil and vegetation than the open plateau tops. However, drought on these islands may induce greater egg and chick mortality through increased collapsing of nest structures.

Studies were conducted on the breeding cycle of Pacific mollymawks on the Chatham Islands in the 1970s and 1990s. Research on Little Sister Island has included a census of breeding pairs, and estimates of breeding productivity and adult survival.

Sooty shearwater *Puffinus griseus* 44 cm

NEW ZEALAND NATIVE, GRADUAL DECLINE

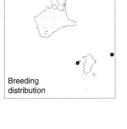

Breeding distribution

Adult sooty shearwater, Rangatira. Photo: Don Merton.

Sooty shearwaters are also known as titi or muttonbirds. They are large black-brown shearwaters with a slender, hooked bill. Their legs and feet are grey-brown with greyish webs. Sooty shearwaters breed on numerous islands in New Zealand, from the Three Kings Islands to Campbell Island as well as the Chatham Islands, where their stronghold is on Tapuaenuku. They also breed on Rangatira, Mangere, Rabbit Island, the Murumurus, Star Keys, The Sisters, Houruakopara, and Kokope. Few birds remain on Chatham or Pitt Islands, as a result of predation by introduced mammals, particularly feral cats and pigs. Young sooty shearwaters were a traditional food source for Moriori and Maori. They are still legally harvested on islands around Stewart Island, but are fully protected elsewhere.

Subantarctic little shearwater *Puffinus assimilis elegans* 30 cm

NEW ZEALAND NATIVE, RANGE RESTRICTED

Also called allied shearwater, this is a small black-and-white shearwater with blue feet and legs. It is slightly larger than a diving petrel, with a more slender bill and body. This subspecies is widely dispersed, breeding in large numbers on the Antipodes Islands, and also on islands in the South Atlantic. The Chatham Islands population is small, with an estimated 100 plus pairs breeding on the Star Keys, and possibly some birds on Tapuaenuku. Little shearwaters are winter breeders, nesting in burrows 0.3–2 metres long.

Breeding distribution

Subantarctic little shearwater near Antipodes Island. Photo: Mike Danzenbaker.

25

Southern diving petrel
Pelecanoides urinatrix chathamensis   20 cm

NEW ZEALAND ENDEMIC SUBSPECIES, NOT THREATENED

Breeding
distribution

Adult southern diving petrel, Rangatira.
Photo: Colin Miskelly.

Southern diving petrels, also called common diving petrels or kuaka, are small stocky seabirds with bright blue legs. The upper plumage is black, with white underneath and greyish under-wings. In the Chatham Islands, they breed on Rangatira, Mangere, Tapuaenuku, the Murumurus, Star Keys, Little Sister Island, Rabbit Island, Houruakopara, and possibly on cliff ledges on Chatham and Pitt Islands. On Rangatira, it has been estimated that there are 164,000 pairs breeding in the forest, but more recent information on the distribution of breeding colonies suggests that there may only be a few tens of thousands of pairs there. Diving petrels are summer breeders; in the Chatham Islands, laying takes place in September–October, in burrows under vegetation.

Snares cape pigeon
Daption capense australe   40 cm

NEW ZEALAND ENDEMIC SUBSPECIES,
RANGE RESTRICTED

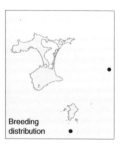

Breeding
distribution

Snares cape pigeon.
Photo: Rod Morris (DOC).

Also known as cape petrel or pintado petrel, the cape pigeon is a medium-sized petrel with distinctive black-and-white check patterning on the back and upper-wings. Noisy flocks often follow fishing vessels at sea. The smaller of the two subspecies breeds in small numbers (tens of pairs) on The Forty Fours and The Pyramid, and in larger numbers on Snares, Bounty, Auckland and Campbell Islands. Cape pigeons are summer breeders on cliff ledges.

Northern giant petrel *Macronectes halli*   90 cm

NEW ZEALAND NATIVE, NOT THREATENED

Breeding
distribution

Giant petrels are also known as nellies, stinkers or stinkpots for their rather unpleasant habit of feeding on carrion such as dead seals and whales. The northern giant petrel is a large robust brown bird with a pale face and throat, and a huge straw-coloured bill with a reddish tip, and prominent nasal tubes. Young birds have entirely dark, almost black, feathers. Southern giant petrels (*M. giganteus*) are generally paler (sometimes all white), with a greenish bill tip. Giant petrels sometimes form noisy flocks squabbling over food behind fishing boats. However, they also catch fish, squid, small seabirds and crustaceans. Northern giant petrels breed on a number of subantarctic islands and in the Chatham Islands, where an estimated 2000 pairs breed on The Sisters and The Forty Fours. They are at risk from fishery by-catch, and also illegal shooting due to their unfavourable reputation.

Northern giant petrels at their nests,
Little Sister Island.
Photo: Rod Morris (DOC).

Adult northern giant petrel, at Rangatira, showing nasal tubes characteristic of all petrels, but most pronounced in giant petrels. Photo: Helen Gummer.

27

Fairy prion *Pachyptila turtur*   25 cm

Breeding
distribution

NEW ZEALAND NATIVE, NOT THREATENED

Prions are medium-sized seabirds with blue-grey upperparts and white underneath. They have a bold black 'M' marking across the back and outstretched wings, and the tail has a broad black tip. The fairy prion is slightly smaller than the broad-billed prion, with similar plumage apart from a wider black tip to the tail. The bill is much smaller and narrower than that of the broad-billed prion. Fairy prions are circumpolar in distribution, and are abundant around New Zealand (more than one million pairs). In the Chatham Islands, c.40,000 fairy prion pairs breed on Mangere, and smaller numbers breed on the Star Keys, Tapuaenuku, the Murumurus, The Sisters, Rabbit Island, and Kokope. They occur at sea in large loose flocks. Fairy prions are vulnerable to rats and other introduced mammals at their breeding grounds.

Adult fairy prion, Mangere, Chatham Islands. Photo: Dave Crouchley (DOC).

Chatham Island fulmar prion
Pachyptila crassirostris pyramidalis

26 cm

CHATHAM ISLANDS ENDEMIC SUBSPECIES,
RANGE RESTRICTED

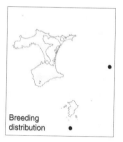

Breeding
distribution •

Identification

The Chatham Island fulmar prion looks similar to the more common fairy prion, but is slightly larger and paler, with a more robust bill. There is some dispute regarding the subspecific status of *pyramidalis*, and the Chatham Island birds are sometimes regarded as the same as those from the large population on the Bounty Islands.

Chatham Island fulmar prion
on nest in rock crevice,
The Pyramid. Photo:
Christopher Robertson.

29

Distribution and ecology

A relatively small population of 1000–5000 pairs breeds in the Chatham Islands, on The Pyramid and The Forty Fours. Another subspecies breeds on the Bounty Islands (c.80,000 pairs) and there are smaller populations at the Snares and Auckland Islands. The fulmar prion's range at sea is poorly known because of the difficulty in identifying it. Some birds remain around their breeding islands all year. Breeding is from November to February, with a single white egg laid in rock crevices, boulder screes, or in burrows in soil under albatross nests.

Threats and conservation

Because fulmar prions are found only on isolated inaccessible islands, they are secure from most human-induced threats. All the islands on which they currently breed are free of introduced mammals; the introduction of rats to any of these islands would be disastrous for prions. There is a need for taxonomic revision of fulmar prions to clarify the relationship between the birds found in the Chatham Islands and other populations.

Chatham Island fulmar prion, The Pyramid. Photo: Christopher Robertson (DOC).

Broad-billed prion *Pachyptila vittata*   28 cm

NEW ZEALAND NATIVE, NOT THREATENED

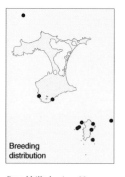

Breeding distribution

Broad-billed prion, Mangere, Chatham Islands. Photo: Dick Veitch (DOC).

Locally known as blue-billies, broad-billed prions have very wide dark grey bills. Larger and darker than other prions, they also have prominent, steep foreheads. Broad-billed prions breed in the South Atlantic and around southern New Zealand. In the Chatham Islands, they breed on Chatham, Pitt, and Rabbit Islands, Rangatira, Mangere, Tapuaenuku, Star Keys, The Sisters, Houruakopara, Kokope, the Murumurus, and Blyth Stack. The largest population is on Rangatira, where there are an estimated 330,000 pairs in forested areas. Broad-billed prions are uncommon in fossil deposits in the Chatham Islands, suggesting that the population has established or increased since human occupation. This has resulted in increased competition for burrows with the endangered Chatham petrel. In recent years Chatham petrel burrows have been actively managed to reduce prion interference. Prions are vulnerable to predation at sites where introduced mammals are present.

Chatham petrel
Pterodroma axillaris

30 cm

CHATHAM ISLANDS ENDEMIC, NATIONALLY ENDANGERED

Other name: ranguru

Identification

The Chatham petrel is a medium-sized grey, white and black petrel. The head, back, tail and upper-wings are slate grey, while the shoulders and upper-wing coverts are a darker grey (giving a dark M pattern across the spread upper-wings). The forehead is mottled grey and white, and the under-parts are white. The under-wing is white except for dark outer tips to the primaries, and a narrow dark trailing edge. A black diagonal band runs from the bend of the wing to the body at the base of the under-wing (this band distinguishes the Chatham Petrel from the similar black-winged petrel, with the latter bird having a narrower band not extending as far as the body).

Breeding
distribution

Chatham petrel, Rangatira.
Photo: Colin Miskelly.

Distribution and ecology

Chatham petrel bones have been found in subfossil deposits of recent age on Mangere, Chatham and Pitt Islands. It appears that Chatham petrels were once widespread over the Chatham group, and were

one of the more abundant burrowing seabirds. By 1900 the breeding range of the Chatham petrel was confined to forested areas on Rangatira, and this remains the sole breeding site. The population is estimated to be about 1000 birds, and 100–130 breeding pairs have been actively managed each year since 1999. Each pair lays a single white egg about December, and chicks fledge in May–June.

Chatham petrel chick, Rangatira, February 2004. Photo: Don Merton.

Threats and conservation

Before European arrival, Chatham petrels are thought to have declined due to predation by kiore (*Rattus exulans*) and muttonbird harvest. The arrival of additional mammalian predators, particularly European rats and cats, and the loss of forest habitat led to the confinement of Chatham petrels to Rangatira. The introduction of mammalian predators to Rangatira, damage to habitat by fire, crushing of burrows by people, and introduction of disease all pose risks to the single population of Chatham petrels. However, the greatest current threat to Chatham petrels is burrow competition from broad-billed prions. While these two species are assumed to have co-existed on Rangatira for a long time, broad-billed prions are now vastly more abundant. As they breed earlier in the year than Chatham petrels, prions also have a competitive advantage when vying for burrows. The greatest impact occurs when prions returning after their moult attack unguarded petrel chicks in their burrows.

Conservation efforts for Chatham petrel began in the late 1980s. The initial focus of this work was locating burrows, and determining the causes of breeding failure. Once it was determined that competition for burrows from broad-billed prions was causing most Chatham petrel breeding attempts to fail, attention shifted to protection of burrows and chicks from prion interference. Chatham petrel burrows are

Underwings of Chatham petrel (above) and black-winged petrel (below), Rangatira. Photos: Don Merton (DOC).

located using radio telemetry—birds that are caught on the surface are tracked to their burrows. Burrows are then protected by installing an artificial, wooden burrow to prevent collapse in the friable soil. Burrows are blocked when the Chatham petrels leave at the end of the season, to ensure that the burrows are not taken over by prions while the petrels are away at sea. During the Chatham petrel breeding season, the burrow is protected using a specially designed neoprene screen that is attached to the burrow entrance and deters prions from entering the burrow. A combination of protection techniques has resulted in a substantial improvement in Chatham petrel productivity at managed burrows. This has made it possible to begin the next stage of the recovery plan: the establishment of a second Chatham petrel population. Up to 55 chicks per year have been transferred to a predator-proof exclosure on Pitt Island, to establish a second population at a site without large numbers of resident prions. Three transfers were successfully completed in 2002–04, and the first birds are expected to return in 2004/05. Further chick transfers are planned to a predator-fenced site under construction on southern Chatham Island.

Black-winged petrel
Pterodroma nigripennis   30 cm

NEW ZEALAND NATIVE, NOT THREATENED

Breeding distribution

Black-winged petrel at Rangatira summit, Chatham Islands. Photo: Colin Miskelly.

The black-winged petrel is a medium-sized grey, white and black petrel. It is distinguished from the Chatham petrel by the black band on the under-wing, which runs along the leading edge of the wing then angles towards the body, but does not reach the "armpit" (unlike Chatham petrel). The black-winged petrel breeds mainly in the Kermadec Islands, but its range is expanding through the southwest Pacific. Small numbers have colonised the Chatham Islands since 1970, and they now breed on Rangatira and Mangere, and probably on Star Keys. Birds have also been recorded prospecting on The Forty Fours and Pitt Island. They are summer breeders and nest in burrows or rock crevices, usually in steep rocky terrain overlooking the sea.

Chatham Island taiko
Pterodroma magentae   38 cm

CHATHAM ISLANDS ENDEMIC, NATIONALLY CRITICAL

Other names: Magenta petrel, tchaik

Breeding
distribution

Adult Chatham Island taiko.
Photo: Graeme Taylor
(DOC).

Identification

Taiko are large gadfly petrels with a sooty-grey head, neck and upper breast, and white underparts. Birds have varying amounts of mottled white around the forehead and chin. Legs and feet are pink with dark outer toes and tips to the webs. They have long narrow wings and are fast, capable fliers. Identification at sea is difficult, and taiko may easily be confused with the soft-plumaged petrel (*Pterodroma mollis*). The name Magenta petrel comes from the ship, the *Magenta*, from which the first taiko to be formally described was collected in 1867.

Distribution and ecology

Taiko disappeared from scientific understanding for over a century, until its rediscovery by David Crockett and his team in the Tuku Valley in 1978. It took another ten years for the first taiko burrow to be discovered near one of the tributaries of the Tuku-a-tamatea River, in southern Chatham Island. Fossil and historic records show that taiko were once the most abundant burrowing seabird on Chatham Island. Oral records described extensive colonies of taiko at the southern

Adult Chatham Island taiko, Taiko Camp. Photo: Gavin Woodward.

end of Chatham Island that were regularly harvested by Moriori. Taiko do not appear to have bred on other islands in the Chatham group; now the species is one of the rarest birds in the world. The entire known taiko population is found in the southwestern forests of Chatham Island, where 160 taiko have been banded between 1978 and 2004. There are 10-15 pairs in known burrows that attempt to breed each year, and the total population is currently estimated at 100-150 birds.

Taiko nest in long burrows under forest cover. They form long-term monogamous pair bonds, and both sexes incubate the single white egg and feed the chick. The breeding season is September–May. During the breeding season, the birds forage over the open ocean, probably to the south and east of the Chatham Islands. Their distribution outside of the breeding season is unknown, although they may disperse eastwards into the temperate South Pacific Ocean. Typical of the genus *Pterodroma*, they specialize in feeding on squid.

Chatham Island taiko chick close to fledging, Tuku Nature Reserve, April 2000. Photo: Graeme Taylor.

Threats and conservation

The arrival of mammalian predators, particularly cats, pigs and rodents, and the introduction of weka, are likely to have been the main causes of the decline of taiko. Loss and degradation of forest habitat as well as the harvesting of the birds for food, particularly once the population was in decline, are also likely to have contributed. The threat to taiko, during the breeding season, from feral cats, pigs, weka and rodents (especially ship rats) continues today. Pig-hunting dogs are also a threat, as are loss of forest habitat from accidental fire, trampling of burrows by feral cattle, and use of burrows as dens by possums.

Since the location of the first burrow in 1987, a number of protection measures have been initiated. Gifting of land for the Tuku Nature

Reserve by the Tuanui family, and the later establishment of a number of adjacent Conservation Covenants by landowners, has given legal protection to key areas of taiko habitat. Fencing of these areas to exclude stock has allowed regeneration and physical protection of the forest. Predator control, targeting feral cats, possums, weka and rats, has been implemented to protect adult taiko visiting the colony, and eggs and chicks in the burrows. The intensive predator control has significantly improved the breeding success at known taiko burrows; since predator control and monitoring of burrows began in 1988, a total of 54 chicks is known to have fledged up to 2004. Known taiko burrows are monitored closely each season to determine the success of protection measures, and to improve understanding of taiko breeding biology and population dynamics.

In order to protect taiko burrows from predators, they must first be located in the dense forest. To do this, radio-tracking operations have been run every 2–3 years. These involve catching taiko as they fly inland to the colony at night, using large spotlights that dazzle the birds and cause them to land on the ground where they can be easily captured. A transmitter is then attached to the bird and it is released back out to sea. Over the following nights it is tracked from strategically placed tracking stations and, once it is tracked going inland, a ground-team heads into the forest to track the bird to its burrow. In more recent years a trained dog has also been used with some success to locate active taiko burrows.

Chatham Island taiko chick, Tuku Nature Reserve, January 1996.
Photo: Graeme Taylor.

There has been a large amount of community involvement in conservation of the taiko, from its initial rediscovery, to the protection of land, to the many volunteers who continue to participate in telemetry and burrow-searches. The Chatham Island Taiko Trust was formed recently to promote taiko conservation work. The Trust has raised funds for the construction of a predator-proof fence to create secure taiko and Chatham petrel breeding colonies. Taiko will be attracted to this safe area using taped petrel calls, and chicks of both species may be translocated to the site so that additional breeding colonies, secure from introduced predators, can be established. Details of the Chatham Island Taiko Trust can be found on their website: www.taiko.org.nz

Grey-backed storm petrel
Oceanites nereis 18 cm

NEW ZEALAND NATIVE, NOT THREATENED

Breeding
distribution

Grey-backed storm petrel,
Rangatira.
Photo: Colin Miskelly.

Storm petrels are the smallest seabirds, and the grey-backed storm petrel is the smallest New Zealand storm petrel. It has a dark grey head, neck, throat and upper breast, and the back, upper-wings and tail are grey, paler towards the rear. The rest of the underparts are white. The tail is square, and the bill and feet are black. At sea it is usually seen alone, darting in the wave troughs. Little is known of numbers. The main New Zealand breeding colonies are on the Chatham, Antipodes and Auckland Islands. There are possibly 10–12,000 pairs on the Chatham Islands, breeding on Rangatira, Mangere, Little Sister Island, The Pyramid, Star Keys, the Murumurus, Rabbit Island, Houruakopara and the stack and islet to the east of Houruakopara. Storm petrels are extremely vulnerable to mammalian predators, and so are restricted to rat-free sites. Grey-backed storm petrels breed over August-March, nesting on the surface under dense vegetation, and occasionally in shallow crevices.

White-faced storm petrel
Pelagodroma marina maoriana   20 cm

NEW ZEALAND ENDEMIC SUBSPECIES,
NOT THREATENED

Breeding
distribution

White-faced storm petrel,
Rangatira.
Photo: Don Merton.

White-faced storm petrels are large storm petrels with white, grey, and dark grey plumage. The face is white with a grey crown, and a grey patch across the eye. They have a distinctive habit of skipping along on the water using their feet. White-faced storm petrels are confined to off-shore islands due to their vulnerability to introduced predators. Despite this, the New Zealand population is large and widespread. The largest population nationally is on Rangatira, where there are an estimated 840,000 breeding pairs in forested areas. They also breed on Tapuaenuku, Little Sister Island, Star Keys, the Murumurus, Rabbit Island, and Kokope, and have recently recolonised Mangere. Adults visit breeding colonies over September–April, and lay a single egg in a burrow about 1 metre long. On Rangatira, white-faced storm petrels are occasionally caught up in vegetation by 'anklets' of the naturally occurring marine trematode worm *Distomum filiferum*. The parasites become attached around the birds' legs, sometimes tangling the legs together and snagging in the trees. In some years this can cause the death of hundreds of birds.

Coastal birds

Maunganui coast, Chatham Island. Photo: Geoff Walls.

Chatham Island oystercatchers on wave platform, Rangatira. Photo: Colin Miskelly.

The Chatham Islands' spectacular coastlines of sweeping sands and rugged basalt cliffs are home to many endemic and threatened bird species. They also provide opportunities to view oceanic species offshore, or to find their remains washed ashore. The two endemic shags are readily seen on headlands, and the oystercatcher is found on many beaches and rocky coastlines, most reliably at Waitangi West, but often near the Waitangi Wharf. Blue penguins breed all around the islands, coming ashore after dark, and leaving their footprints across sandy beaches. Brown skuas frequently visit main Chatham, but are more easily seen on Pitt Island and around outlying islands. Shore plover can only be seen from boats close offshore Rangatira and Mangere, while migrant waders can be found along the margins of Te Whanga Lagoon, especially on the northeastern shore near Hapupu.

Chatham Island blue penguin
Eudyptula minor chathamensis 40 cm

Other names: korora, fairy penguin, little penguin

Identification
Blue penguins are the smallest of all penguin species. They have slate-blue upperparts with white below. Sexes are alike although males have longer and deeper bills. Six subspecies of blue penguin have been described, but further taxonomic work is required. Blue penguins are mostly nocturnal on land, and are often heard calling noisily from their nest sites at night.

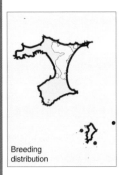

Breeding
distribution

Adult Chatham Island blue penguin, Star Keys. Photo: Adam Bester.

Pair of Chatham Island blue
penguins, Rangatira.
Photo: Graeme Taylor.

Distribution and ecology

Blue penguins breed on Chatham Island, Pitt Island, Rangatira, Mangere, Star Keys, Houruakopara, and Kokope. Whilst there are large numbers breeding on Rangatira and Star Keys, the populations on the main islands may be declining due to feral cat predation. There are an estimated 5000–10,000 pairs of Chatham Island blue penguins. Blue penguins spend most of the year at sea, but come to shore to breed and moult. Nests are constructed in burrows, natural cavities or under rocks or buildings. Many birds nest close to the shore, but others will travel hundreds of metres inland and up hill. Blue penguins can nest in loose colonies or as isolated pairs.

Threats and conservation

The major threats to blue penguins on land are dogs, feral cats and pigs, which kill adult penguins and chicks if they have access to them. Norway rats may take eggs and young chicks on Chatham Island, but other rodents do not appear to be a threat. On Chatham and Pitt Islands, weka may also take eggs and young chicks. Cattle and sheep trample some nests, although most are protected in rock crevices or hollows in tree trunks. Fire is a risk to penguins, particularly when they are nesting and moulting.

The gazettal of Rangatira and Mangere as Nature Reserves, the removal of stock, and ongoing rodent quarantine measures, protect blue penguin populations on these two islands. The fencing of a number of coastal reserves and covenants on Chatham and Pitt Islands has also provided more secure nesting habitat for penguins. Penguins appear to have benefited from predator control undertaken to protect Chatham Island oystercatchers on northern Chatham Island beaches, but this has not been studied in any detail.

Chatham Island shag
Leucocarbo onslowi   63 cm

CHATHAM ISLANDS ENDEMIC, RANGE RESTRICTED

Identification

The Chatham Island shag is a large shag with a blue eye-ring, orange fleshy growths above the bill, and orange facial skin below the bill. Plumage is mostly iridescent black, with white on the breast and belly, and white patches on the upper wing. The feet are pink. Sexes are alike. The facial skin colour fades outside the breeding season. In flight, the head is held lower than the body.

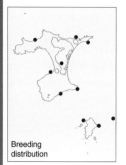

Breeding
distribution

Adult Chatham Island shag,
Star Keys.
Photo: John Kendrick
(DOC).

42

Distribution and ecology

Chatham Island shag breeding colonies are located on headlands or stacks off the coast of Chatham Island, in Te Whanga Lagoon, and on Rabbit Island, North East Reef and Star Keys. In a census of all shags in the Chatham Islands in 1997, 842 breeding pairs of Chatham Island shag were counted at ten discrete breeding sites. However, a census in 2003 found only 270 pairs at the same sites; it is not yet known whether this is due to natural variation or a serious decline.

Chatham Island shags typically feed offshore, but occasionally do so near the shoreline. They usually feed alone but often roost in flocks of 50 or more birds, and breed in colonies. Nests are made of seaweed and other vegetation cemented with guano. Up to four pale blue eggs are laid over August–December.

Chatham Island shag colony, Cape Fournier, January 1978. Photo: Reg Cotter.

Threats and conservation

The largest colonies of Chatham Island shags occur at predator-free sites, but some colonies are still present on Chatham Island and are subject to disturbance by feral cats, weka, possums, pigs, sheep, cattle, dogs and people (including illegal shooting). Disturbance by people can lead to birds stampeding from nests, causing breakage of eggs, or subsequent egg or chick predation by gulls. Fur seals disturb nesting at the Star Keys, and have occupied former colony sites there.

No Chatham Island shag colonies currently have legal protection, although a number of colonies have some measure of protection by virtue of their inaccessibility and isolation. Further surveys of breeding sites, and counts of breeding pairs, are required to determine whether the huge decrease in breeding pairs between 1997 and 2003 was due to a single poor breeding season, or was caused by a population decline.

Chatham Island shag colony, Cape Fournier, January 1979. Photo: Colin Miskelly.

43

Pitt Island shag
Stictocarbo featherstoni　　　　63 cm

CHATHAM ISLANDS ENDEMIC, RANGE RESTRICTED

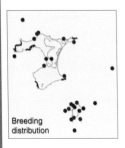

Breeding
distribution

Pitt Island shag on
nest, Tapuaenuku,
September 1976.
Photo: Dick Veitch (DOC).

Identification

The Pitt Island shag is more slender than the Chatham Island shag, and has yellow feet, a long slender bill, a dark head, double crest and upperparts, and light grey underparts. The wings are dark grey with small black spots, similar to those of the closely related spotted shag around coasts of mainland New Zealand. The lime-green face and white plumes at the back of the neck are particularly prominent just before breeding. Sexes are alike. Pitt Island shags can be recognised in flight by their slim silhouette, dark plumage, and because they hold their neck and head horizontal.

Distribution and ecology

The Pitt Island shag is a marine species restricted to the Chatham Islands. It breeds in small colonies of 3-20 pairs on rocky headlands, coastal cliff ledges and islets. Pitt Island shags breed on Chatham, Pitt, and Rabbit Islands, Rangatira, Mangere, Star Keys, Tapuaenuku, The

Castle, some of the Murumurus, The Pyramid, The Forty Fours, The Sisters, and Western Reef. Some also breed on Shag Rock (off Motuhinahina) in Te Whanga Lagoon. The total Pitt Island shag population is estimated to be less than 700 breeding pairs. Nests are platforms of seaweed and other vegetation cemented with guano. Three pale blue eggs are laid during August to December. Nest sites are often changed between years.

Threats and conservation

Pitt Island shags usually nest at sites inaccessible to dogs, pigs or stock. However, a few nests on Chatham and Pitt Islands are at sites accessible to feral cats and weka. Some eggs and chicks may be lost to rats on Chatham Island. Pitt Island shags are sometimes caught in crayfish pots and cod pots, and illegal shooting has occasionally been reported.

Colonies on Rangatira and Mangere are protected by the gazettal of the islands as Nature Reserves, the removal of stock, and quarantine measures implemented to prevent the arrival of mammalian predators. A survey of breeding sites and census of breeding pairs was conducted by the Department of Conservation in 1997 and 2003, revealing an apparent decline of 25% over the 6 years.

Top: Pitt Island shag in breeding plumage, with Mangere in background, October 1976. Photo: Rod Morris (DOC)
Above: Juvenile Pitt Island shags, Rangatira, December 1983. Photo: Colin Miskelly..

White-faced heron *Ardea novaehollandiae* 67 cm

NEW ZEALAND NATIVE, NOT THREATENED

Sometimes called the blue heron, the white-faced heron is a slender blue-grey bird with white around the face. It has a sharp black bill and long yellow legs. White-faced herons are generalist feeders, and thrive in many habitats including rocky shores, lake margins, damp pasture, and swamps. They have benefited from the widespread conversion of forest to farmland throughout New Zealand, and colonised the Chatham Islands from mainland New Zealand in the early 1970s. They breed in pairs or loose groups, building large untidy nests high up in macrocarpas or similar trees, or, in the Chatham Islands, on cliff ledges or in sea caves. They have been recorded breeding on Chatham Island, Rangatira, and Mangere.

White-faced heron.
Photo: Peter Reese.

46

Chatham Island oystercatcher
Haematopus chathamensis   48 cm

CHATHAM ISLANDS ENDEMIC, NATIONALLY CRITICAL

Other names: torea, Chatham Island pied oystercatcher

Chatham Island
oystercatcher, Rangatira.
Photo: Don Merton.

Identification

The Chatham Island oystercatcher is a large black-and-white wader with a long red bill and sturdy pink legs. It is most similar to the pied phase of the variable oystercatcher (*H. unicolor*) of the New Zealand mainland. Females have longer, thinner bills and are slightly larger than males. The similar but smaller New Zealand pied oystercatcher (*H. finschi*) occasionally visits the Chatham Islands, but has not been recorded interacting with Chatham Island oystercatchers.

Distribution and ecology

The Chatham Island oystercatcher is the rarest oystercatcher in the world. It is found on Chatham and Pitt Islands, Rangatira, and Mangere, and occasionally the Star Keys, on both rocky coastlines and sandy beaches. In the southern part of its range (Rangatira, Mangere, and Pitt Island) and in southern Chatham Island their habitat is dominated by rocky habitats with extensive wave platforms. In the northern part of the range they use a mix of sandy beaches and rocky wave platforms, especially near stream mouths.

Above: Chatham Island oystercatcher nest in an old car tyre placed as a nest platform, north coast Chatham Island, October 2001. Photo: Reg Cotter.

Top right: Chatham Island oystercatcher nest, north coast Chatham Island, November 2003. Right: Chatham Island oystercatcher chicks hiding in tide-wrack bull kelp, north coast Chatham Island, November 2003. Photos: Colin Miskelly.

Oystercatcher numbers have increased substantially in recent years in response to management, particularly in northern Chatham Island. Since 1987, the breeding population has increased from 42 to 85 pairs, and the total population has increased from fewer than 150 to around 290 birds by 2004, with the bulk of the population on north Chatham Island beaches. They form monogamous long-term pair bonds, and are strongly territorial during the breeding season; many pairs stay attached to their breeding territories year round. Eggs (2–3) are well camouflaged, and laid in a simple scrape in the sand, or in a depression or small crevice among rocks or driftwood, during October–January.

Threats and conservation

The key threat to oystercatchers on Chatham Island is predation by feral cats. They are the main predator of oystercatcher eggs, and also prey on chicks and adults. Weka and red-billed gulls take some oystercatcher eggs, and trampling by stock is a threat to nests. Disturbance during nesting by stock, people and dogs wandering along the coast causes birds to leave the nest, increasing the risk of eggs becoming chilled or overheating, and exposing them to predators. Crushing of eggs by quad bikes and vehicles used to launch dive boats is also an ongoing threat at some sites, particularly Waitangi West.

Changes in coastal vegetation, and the establishment of introduced marram grass, appear to have had an adverse effect on oystercatcher breeding. Marram stabilises the dunes, causing the beach profile to become steeper. This reduces the area of suitable beach available for oystercatchers to nest on, forcing the birds to nest further down the beach profile. There the likelihood of losing nests to high tides or storm surges is greater.

Chatham Island oystercatchers received little scientific attention until the 1970s, and management was not attempted until the 1990s. There was sporadic predator control and management in northern Chatham Island in the mid 1990s, resulting in 1–11 chicks being produced per year from 10–14 pairs. Research, focused on the breeding biology and habitat use of the oystercatcher population on northern Chatham Island, was completed in 1998, following which more intensive management was initiated. This was coupled with a further research programme monitoring colour-banded birds, and using video surveillance to assess the effectiveness of management, to establish the causes of nest failure, and to identify key predators. Intensive management is continuing, and research is to continue into oystercatcher population trends and dynamics to guide recovery effort. Management involves trapping key predators, particularly feral cats, and protecting nests by fencing off sections of beach. Where this is not possible, individual nests are fenced to keep stock away. The problem of high seas washing away nests is tackled in two ways. Firstly, vulnerable nests are gradually moved up the beach, away from the waves, without disturbing the birds. In some instances, the birds are encouraged to nest in old car tyres attached to boards which can be dragged higher up the beach. The longer-term solution is dune restoration, where marram grass is removed and native vegetation restored to create a more open dune environment, where oystercatchers can nest further back from the shore with less risk of predation. Application of these techniques to protect oystercatchers along the coast between Maunganui and Wharekauri has resulted in 25–35 chicks fledging each year, with breeding success about 3 times greater than at unmanaged sites.

New Zealand shore plover
Thinornis novaeseelandiae   20 cm

NEW ZEALAND ENDEMIC, NATIONALLY CRITICAL

Other name: tuturuatu

Adult male New Zealand
shore plover, Rangatira.
Photo: Don Merton.

Identification

Shore plovers are small, stocky birds with white, black and brown plumage. They have a brown cap, with a white ring separating this from the dark face and neck. The face is black in the male and dark brown in the female. The bill is red with a black tip, which is more extensive and less sharply defined in the female; the legs are orange.

Adult female New Zealand
shore plover, Rangatira.
Photo: Don Merton.

Distribution and ecology

Shore plover were once widespread in coastal areas of New Zealand, but by the late 1800s they became confined to the Chatham Islands. There is currently only one natural wild population on Rangatira, where there are about 130 birds, comprising about 45 breeding pairs. Recent transfers from Rangatira to Mangere have led to the establishment of a small population there.

The discovery of a tiny shore plover population on Western Reef in 1999 came as a surprise to all those involved with the species.

Juvenile New Zealand shore plover, Rangatira, February 2004. Photo: Don Merton.

Unfortunately this tiny population (21 birds at the highest count) declined steadily after its discovery to only one male in 2003. The last male was captured in June 2003 and taken into captivity at the National Wildlife Centre at Mount Bruce, north of Masterton, where it has mated with a captive female (ex Rangatira stock) already held there, and produced a single male chick in 2003/04. Progeny, with their unique Western Reef genetic material, will be returned to the Chatham Islands for establishment of another wild population.

Habitat use by shore plover appears to be flexible. On Rangatira they use the rocky coastline, wave platforms, salt meadows and freshwater seeps. However, historically shore plover were recorded in estuarine and sandy habitats on the New Zealand mainland. In recent releases of captive-bred shore plover, the birds have often favoured sandy habitats. Shore plover feed on a wide range of small aquatic and terrestrial invertebrates. Bulky nests are built under the cover of vegetation, logs or boulders. The breeding season is September–February (occasionally to April); laying (2–3 eggs) peaks in October.

Threats and conservation

New Zealand shore plover nest with newly hatched chicks, Rangatira, December 1983. Photo: Colin Miskelly.

The retreat of shore plover from their traditional range coincided with the spread of cats and Norway rats during the 1800s. Shore plover disappeared from Mangere and Pitt Island after the introduction of cats. The cause of the more recent decline of the tiny shore plover population on Western Reef is not clear, but it followed a rapid increase in the reef's fur seal population. Current threats to shore plover survival on Rangatira are the introduction of predators or disease, or habitat modification such as fire. Forest and vine regeneration of former farmland on Rangatira has caused some reduction in breeding habitat for shore plover; this process is continuing, although now threatens only a small proportion of the population.

The recovery programme for shore plover aims to protect and extend the species' range within the Chatham Islands, and to re-establish populations in other parts of New Zealand. To facilitate this, a captive population was set up on mainland New Zealand to breed birds to found new wild populations. To date, a population has been established on an island off the east coast of the North Island. Transfers have also been undertaken within the Chatham Islands. Transfers from Rangatira to Mangere in the early 1970s failed to establish a population. However, three transfers conducted in 2001–03 were successful, with several pairs now breeding on Mangere. Further transfers are planned.

Brown skua *Catharacta skua lonnbergi*   63 cm

NEW ZEALAND NATIVE, SPARSE

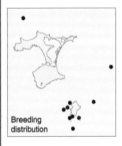

Breeding
distribution

Also known as southern skua, subantarctic skua, sea hawk, or hakoakoa, this is a large, dark brown gull-like bird with a large, hooked black bill. The wings are broad, with white flashes at the base of the primaries. Females are heavier than males. The brown skua breeds on numerous islands in the subantartic and antarctic region. It is considered a full species by some experts. There are an estimated 200 skuas in the Chatham Islands, most of which breed on Rangatira and Mangere. Smaller numbers breed on the

Pair of brown skuas,
Rangatira.
Photo: Colin Miskelly.

Star Keys, Tapuaenuku, Rabbit Island, the Murumurus, southern Pitt Island, The Sisters, The Forty Fours, The Pyramid, and The Castle, and they range to the rest of Pitt Island and coasts of Chatham Island. Brown skuas are strongly territorial when breeding, and will dive aggressively at people passing too close to their nest. While most skuas breed in pairs, a significant number breed in trios comprising two unrelated males and one female; breeding groups of up to seven birds have also been recorded. Skuas eat small seabirds such as prions,

storm petrels and diving petrels, as well as fish, goose barnacles, eggs and scavenged carcasses. Skuas are sometimes shot on Chatham and Pitt Islands, where they occasionally attack cast sheep or lambs. This is of concern, as the brown skua is one of the rarest bird species on the Chatham Islands.

Brown skua diving to defend its nest,
Mangere, December 1982.
Photo: Dave Crouchley (DOC).

Southern black-backed gull
Larus dominicanus

60 cm

NEW ZEALAND NATIVE, NOT THREATENED

Southern black-backed gull feeding chicks at nest. Photo: Peter Morrison (DOC).

The southern black-backed gull, also known as the Dominican gull, kelp gull, or karoro, is a large black-and-white gull with a yellow bill and greenish legs. The bill has an orange spot at the tip of the lower mandible. Juvenile plumage is dark brown in the first year, becoming mottled buff-and-white in the second year, gradually changing to adult plumage by the fourth year. Sexes are alike. Black-backed

Immature southern black-backed gull. Photo: Fred Kinsky (DOC).

gulls inhabit coasts, lakes and open country, and have adapted extremely well to modified environments, readily utilising new food sources such as rubbish dumps and fishing waste. They are abundant on the Chatham Islands, breeding on Chatham, Pitt, and Rabbit Islands, Rangatira, Star Keys, The Sisters, The Forty Fours, The Pyramid, Houruakopara, Kokope, and Blyth Stack. Breeding occurs during October–February, usually in colonies, although some solitary nests are found on coastal rock stacks and headlands. The only estimate of numbers in the Chatham Islands has been on Rangatira, where 20–41 active nests were counted annually between 1986 and 1995, but there are larger colonies on Chatham Island.

Red-billed gull
Larus novaehollandiae scopulinus 37 cm

NEW ZEALAND ENDEMIC SUBSPECIES, NOT THREATENED

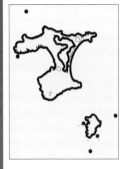

Red-billed gull on nest, Rangatira, December 1981. Photo: Dave Crouchley (DOC).

The red-billed gull is a medium-sized, predominantly white-and-grey gull with distinctive red bill, legs and eyelids. The head, breast, underparts and tail are pure white; the back and wing coverts are pale grey and the primaries are black with white spots. Immature red-billed gulls have dark bills, eyes and legs, and dark speckles on the upper-wing in their first plumage. Red-billed gulls are found in coastal areas throughout New Zealand, including most islands in the Chatham Islands group, where they breed on Chatham Island (including Te Whanga Lagoon), Rangatira, Mangere, Star Keys, Houruakopara, Little Sister Island, The Pyramid, Rabbit Island, Kokope, Western Reef, and Blyth Stack. Red-billed gulls are gregarious, roosting and feeding in large flocks, and breeding in densely packed colonies. Breeding extends from July to January. They have benefited from human settlement through increased availability of food in the form of fishing and other waste. There are no comprehensive estimates of numbers of red-billed gulls for all of the Chatham Islands. The maximum combined number of nests on Rangatira and Mangere in any one season between 1986 and 1995 was 54.

White-fronted tern *Sterna striata* 42 cm

NEW ZEALAND NATIVE, GRADUAL DECLINE

White-fronted tern on nest,
Rangatira, November 1975.
Photo: Rod Morris (DOC).

The white-fronted tern or tara is a medium-sized tern with a long black bill. Plumage is pale pearly grey above and white below, with a black cap separated from the bill by a white forehead (the "front"). There are estimated to be several hundred pairs in the Chatham Islands, breeding on steep cliffs and rocky islets off Chatham Island, around Te Whanga Lagoon, and on Rangatira, Mangere, Star Keys, The Sisters, The Forty Fours, The Pyramid, the Murumurus, and Houruakopara. Most of the Chatham population disappears in autumn, and the birds are thought to migrate to New Zealand, or to southeastern Australia (where there has been one recovery of a tern banded on the Chatham Islands), along with many from the New Zealand population. White-fronted terns are gregarious, breeding in colonies and feeding in flocks, often with other species; they hover and dive on shoaling fish.

White-fronted tern colony, The Pyramid,
September 1974.
Photo: Christopher Robertson (DOC).

55

Migrant waders

Pacific golden plover *Pluvialis fulva*   25 cm

This mottled brown-and-golden-buff plover is a regular summer visitor, in small numbers, to Te Whanga Lagoon, especially near Hapupu. It has a comparatively large head, a short black bill and long dark legs. Adults often develop the black face and underparts of breeding plumage before leaving New Zealand for their Siberian and Alaskan breeding grounds in April.

Pacific golden plover in non-breeding plumage.
Photo: Peter Reese.

Turnstone *Arenaria interpres*   23 cm

Also called ruddy turnstone, this is a small stocky wader with orange legs, white underparts and tortoise-shell plumage above. During the non-breeding season the plumage on the upper body is a duller brown. Turnstones breed on Arctic coasts and migrate to the southern hemisphere for the northern autumn and winter. Over 500 visit the Chatham Islands during October–April, mainly occurring on the north coast of Chatham Island, and on the northern and eastern margins of Te Whanga Lagoon.

Turnstones in non-breeding plumage.
Photo: Peter Reese.

Lesser knot *Calidris canutus* 24 cm

The lesser knot is a plump, short-legged, grey-and-white wader with a heavy, straight black bill. The head, neck and breast become rusty red when in breeding plumage. The lesser knots that migrate to New Zealand (including the Chatham Islands) breed at high latitudes in northeast Siberia. Up to 1800 knots visit Chatham Island during the summer months, occurring in small flocks around Te Whanga Lagoon, especially along the eastern shore.

Lesser knots in non-breeding plumage, with one in partial breeding plumage.
Photo: Brian Chudleigh.

Eastern bar-tailed godwit
Limosa lapponica baueri 40 cm

This large brown-and-grey wader is also known as kuaka. It has long legs and a long, slightly up-curved bill. Females are bigger than males, and have longer bills. Birds often begin to show red breeding plumage towards the end of our summer, before departure from New Zealand to their breeding grounds in Alaska and northeast Siberia. At least 300 godwits regularly visit the Chatham Islands over the summer months, mainly around the eastern margins of Te Whanga Lagoon.

Eastern bar-tailed godwits in non-breeding plumage.
Photo: Dick Veitch (DOC).

Freshwater birds

Te Whanga Lagoon.
Photo: Jeremy Rolfe.

There are more than 50 lakes and lagoons on the Chatham Islands, including the enormous brackish Te Whanga Lagoon (18,600 ha), seemingly providing abundant habitat for waterfowl. But this group of Chatham Island birds has been most hard-hit by human-induced changes. At least nine species are globally or locally extinct, including a large endemic flightless duck, an endemic shelduck, a merganser, and the New Zealand little bittern. The remaining species are all common on the New Zealand mainland, but often lakes on the Chathams appear surprisingly devoid of birdlife. The shores of Te Whanga Lagoon provide the best opportunities for viewing.

Tennants Lake, Chatham
Island. Photo: Jeremy Rolfe.

Black shag *Phalacrocorax carbo* 88 cm

NEW ZEALAND NATIVE, SPARSE

Breeding
distribution

Black shags are also known as great, common or black cormorants, or kawau. They are the largest of the shags, and have black plumage that is somewhat browner on the wings and tail, yellow facial skin with a white band on the cheeks, and a white patch on the flank when breeding. Black shags nest in small colonies in tree-tops or among flax clumps. A survey conducted in 1997 found five black shag colonies, all on Chatham Island, with a total of 233 breeding pairs. Black shags feed on small and medium-sized fish in lakes, rivers and sheltered inshore coastal waters.

Black shag at nest.
Photo: Peter Reese.

Black swan *Cygnus atratus* 120 cm

RE-INTRODUCED TO NEW ZEALAND AND CHATHAMS

Black swan family.
Photo: (DOC).

Black swans are very large, black, long-necked birds, with white wing-tips. The crimson bill has a white bar at the tip. Sexes are alike. Native to Australia and New Zealand, black swans became extinct in New Zealand and the Chatham Islands before European arrival. About 100 black swans were re-introduced to New Zealand during the 1860s; more are thought to have arrived naturally, and they became well established throughout New Zealand. Four or five were brought to Chatham Island by Walter Hood in 1890, and others are likely to have flown from New Zealand or Australia. A large

Black swan in flight.
Photo: John Kendrick
(DOC).

population is now centred on Te Whanga Lagoon, and they are also found on other lakes on Chatham and Pitt Islands. Numbers fluctuate depending on seasonal conditions. In particular, high water levels in the lagoon can restrict birds feeding on the vegetation on the lakebed and cause large die-offs. Estimates have varied from as high as 30,000 in 1953 to around 3000 birds in 1981. Timing of breeding varies in relation to local conditions. Nests are large mounds of grass, usually built within 100 m of a lake. The large pale green eggs are often harvested for food in the Chatham Islands. During the 1950s, it was estimated that up to 40,000 eggs were collected annually.

Feral goose *Anser anser* 80 cm

INTRODUCED TO NEW ZEALAND AND CHATHAMS

Feral geese are the familiar large white (male), brown-and-white (female), or grey-brown (juvenile) domestic geese of farm ponds throughout New Zealand and the Chatham Islands. Feral geese occur at several locations

Feral goose family. Photo: Peter Reese.

on Chatham and Pitt Islands, often grazing on pasture away from water.

Grey duck *Anas superciliosa* 55 cm

NEW ZEALAND NATIVE, SERIOUS DECLINE

The plumage of both sexes of grey duck is similar to that of a female mallard, but grey ducks are slightly darker and have a pale head with a conspicuous dark eye-stripe and cap. They have a glossy green speculum on the upper-wing, which is visible in flight and sometimes when preening. Grey ducks interbreed with mallards, and many birds are hybrids. Grey ducks prefer freshwater habitats such as small lakes and slow-flowing rivers although they are sometime seen in tidal areas. They tend to avoid waters surrounded by farmland or close to human habitation. Grey ducks breed on Chatham and Pitt Islands and occasionally on Rangatira. Numbers have declined with the increase in introduced predators, shooting, and establishment of black swans. Grey ducks in New Zealand and the Chatham Islands are now severely threatened through hybridisation with introduced mallards.

Grey duck.
Photo: Jeremy Rolfe.

61

Mallard *Anas platyrhynchos* 58 cm

INTRODUCED TO NEW ZEALAND

Males in breeding plumage have a glossy green head, chestnut breast, grey body and black rump. The female is speckled buff and dark-brown. The male in non-breeding plumage is similar to the female, but often with a few remnants of breeding colours. Both sexes have a glossy blue speculum, bordered with black-and-white stripes, on the upper-wing. Plumage is variable

Mallard drake. Photo: Jeremy Rolfe.

because of the high level of hybridisation with grey ducks. Found naturally throughout the temperate Northern Hemisphere, mallards were introduced to New Zealand from the 1860s through to the 1960s, and reached the Chatham Islands about 1950. Mallards are now the most numerous waterfowl in New Zealand and are found on almost any type of fresh or brackish water. They are common on Chatham Island and are regularly seen on Pitt Island and Rangatira.

Pukeko *Porphyrio melanotus* 51 cm

NEW ZEALAND NATIVE, NOT THREATENED

Pukeko (swamphens) are large rails with sooty-black necks, back and upper-tail, and with a purplish-blue throat and breast. They have a prominent red bill and frontal shield, and long, reddish legs with long

spreading toes. The under-tail is white and is conspicuous as it is flicked up with each step. Pukeko inhabit lowland swamps and rough farmland, and have benefited from land clearance on mainland New Zealand, Chatham, and Pitt Islands.

Pukeko feeding.
Photo: Rod Morris (DOC).

62

Spotless crake *Porzana tabuensis*   20 cm

NEW ZEALAND NATIVE, SPARSE

Spotless crakes are tiny sooty-coloured rails with a chestnut back, and a bright red eye and eye-ring. The under-tail is barred black and white. Crakes are rarely seen but occasionally heard, particularly in response to taped 'mook' and 'purrrrr' calls. They are distributed throughout New Zealand, including the Chatham Islands, mostly in dense sedge-dominated swamps. Estimates of numbers and distribution are difficult because of their secretive nature.

Spotless crake on nest.
Photo: Geoff Moon (DOC).

Marsh crake *Porzana pusilla*   18 cm

NEW ZEALAND NATIVE, SPARSE

Marsh crakes are tiny rails with brown upperparts spotted with white and streaked with black. The breast is grey and the abdomen barred black and white. They occur in swamps and thick reedy lake margins on mainland New Zealand. Rarely seen and quieter than spotless crakes, most records are of cat-killed birds or birds flushed out of vegetation by dogs. They were last reported on Chatham Island in the early 1900s, but may still be present.

Marsh crake.
Photo: Peter Moore (DOC).

Australasian pied stilt
Himantopus himantopus leucocephalus   35 cm

NEW ZEALAND NATIVE, NOT THREATENED

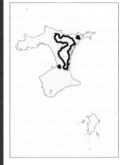

The pied stilt is a slim black-and-white wader with very long red legs, a long, fine black bill, and a dark red eye. The back and wings are black, and the face and underparts are white. There are varying amounts of black on the neck and crown. Juvenile plumage is mottled with a grey wash on the head and neck, and brown-black on the back and wings. Stilts have a distinctive yapping call; the long legs and feet trail behind in flight. Pied stilts were first recorded breeding on Chatham Island in 1961, but are not common there. They have been recorded only from the main island, mainly around the margins of Te Whanga Lagoon, but also at other lakes. Breeding is from July to January; stilts nest as isolated pairs or loose groups, generally surrounded by, or near, water.

Pied stilt at nest. Photo: Barry Harcourt (DOC).

Birds of open country

Farmland on Pitt Island, with Rangatira in background. Photo: Ron Nilsson (DOC).

Most of Chatham and Pitt Islands has been cleared for farming, but large areas have subsequently become overgrown with bracken fern, introduced gorse, and the distinctive "Clears" vegetation dominated by low shrubs and bamboo rush. Some areas are regularly burnt to remove woody vegetation, maintaining open habitats that are little used by native birds. The commonest species here are all European species introduced to mainland New Zealand, and which subsequently flew to (or were introduced to) the Chatham Islands. Notable exceptions are the endemic pipit (also common along coasts), the Australasian harrier, and the buff weka. Weka are not native to the Chatham Islands, but were introduced from Canterbury in 1905. On Chatham Island, they are abundant from the shoreline to the forest interior, and are most easily seen foraging in damp pasture on overcast days.

Australasian harrier *Circus approximans*   55 cm

NEW ZEALAND NATIVE, NOT THREATENED

Australasian harrier.
Photo: Peter Reese.

Also known as swamp harrier, kahu or hawk, this is the only raptor currently known from the Chatham Islands. The harrier is a large hawk with long-fingered wings forming a shallow V in flight, and a long, slightly rounded tail. It has dark brown upperparts with buff underneath. Harriers become paler with age; females are larger than males. They are frequently seen soaring over open farmland, swamps and forest edges, in search of carrion or prey. Australasian harriers are widespread throughout Australasia and have become more numerous in New Zealand, and the Chatham Islands, as a result of land clearance for farming. They are common and breed on Chatham and Pitt Islands, are usually present on Rangatira, and less often on Mangere.

California quail *Callipepla californica*   25 cm

INTRODUCED TO NEW ZEALAND AND CHATHAMS

The California quail is a large brown, white and grey quail with a forward-curving plume on the head. The male has a black throat with a white band under it. This North American species was introduced to New Zealand in the 1860s, and is presumed to have been introduced to the Chatham Islands before 1950, where it may survive in small numbers on Chatham and Pitt Islands. California quail prefer a mix of rough pasture and low scrub.

Male California quail.
Photo: Peter Reese.

Buff weka *Gallirallus australis hectori*   53 cm

NEW ZEALAND ENDEMIC, INTRODUCED TO CHATHAMS

Weka are hen-sized, flightless rails, with buff-brown plumage and sturdy bills and feet. They are endemic to mainland New Zealand. The buff weka, a subspecies formerly found in the eastern South Island, was introduced to Chatham Island in 1905: 12 birds were released at Kaingaroa by Walter Hood. They disappeared from their natural range by the 1930s but are abundant on Chatham Island, from where they were introduced to Pitt Island in the 1970s. Historically, weka provided an important food source to both Maori and Pakeha in many parts of New Zealand. Weka are fully protected elsewhere in New Zealand, but can be legally harvested in the Chatham Islands where weka-hunting is a popular activity.

Weka occupy a wide variety of habitats including coastal areas, rough pasture, forests and wetlands. Their diet consists mostly of invertebrates and fruit, but they also feed on rodents, the eggs and young of ground-nesting birds, and carrion. Weka

Adult buff weka, Chatham Island. Photo: Peter Reese.

are generally territorial and can breed all year round if food is abundant, laying up to six eggs per clutch.

The decline of weka in their historic range has been linked to a variety of factors including climate, food supply, disease, changes in habitat and the introduction of mammalian predators, particularly ferrets and stoats. On the Chatham Islands, large numbers of weka are harvested each year for food, while others are killed by dogs and, at some sites, as part of protection programmes for the endangered taiko and Chatham Island oystercatcher. While there is no information regarding the impact of this take on the population, weka appear to be flourishing on both Chatham and Pitt Islands and may number over 50,000 birds.

There have been several attempts to transfer buff weka back to their natural range in the South Island. A reintroduction to Arthurs Pass in 1962 was not successful, but a transfer in 2002 to Te Peka Karara, an island in Lake Wanaka, is showing promising results.

Banded dotterel
Charadrius bicinctus bicinctus   20 cm

NEW ZEALAND NATIVE, GRADUAL DECLINE

Banded dotterels are small plovers with grey-brown upperparts; they are white underneath except for two bands across the neck and breast, which vary depending on age, sex and time of year. Breeding adults have a thin black band on the upper breast and a broad chestnut band below this, which is broader and darker in the male. Banded dotterels of the Chatham Islands are assumed to be the same subspecies as those on the North and South Islands. They breed on Chatham and Pitt Islands, and occasionally on Rangatira. Banded dotterels are found in areas of low vegetation known as 'Clears' on both Chatham and Rangatira, and also around Te Whanga Lagoon, in swampy pasture and in coastal areas. The total population in the Chatham Islands is estimated to be in the order of 200–300 birds. The banded dotterel population in New Zealand is partially migratory, with many birds migrating to northern New Zealand or Australia after the breeding season. However, banded dotterels on the Chatham Islands are apparently non-migratory. During winter, dotterels form small flocks, then from August onwards, they defend separate territories. Nests are usually built in open areas, providing the incubating bird a clear all-round view.

Male banded dotterel
on nest.
Photo: Dick Veitch (DOC).

Spur-winged plover
Vanellus miles novaehollandiae

38 cm

NEW ZEALAND NATIVE, NOT THREATENED

Also known as masked lapwing, this is a large olive-brown, white and black plover with distinctive yellow wattles on the face, and a sharp spur protruding from the carpal joint on the wing. Self-introduced to New Zealand from Australia in the 1930s, the spur-winged plover has spread widely through New Zealand. Breeding was first recorded on Chatham Island in 1981, and they have since become well established there and on Pitt Island. Adults aggressively defend their nests and young, and are often seen chasing harriers.

Spur-winged plover on nest.
Photo: (DOC).

Welcome swallow
Hirundo tahitica neoxena

15 cm

NEW ZEALAND NATIVE, NOT THREATENED

The welcome swallow is a recent colonist in New Zealand, and has bred on Chatham Island since about 1976. It is a small, slender bird with rapid, darting flight. Welcome swallows have dark heads and backs, chestnut-coloured foreheads and throats, and are greyish underneath. The wings are long and pointed and the tail deeply forked. Welcome swallows are common near waterways and in open areas. They forage by hawking for flying insects. The characteristic mud nests are attached to vertical surfaces under an overhang, typically under eaves of buildings, bridges and culverts, or in sea caves.

Welcome swallows at their
nest. Photo: Peter Reese.

Chatham Island pipit
Anthus novaeseelandiae chathamensis   19 cm

CHATHAM ISLANDS ENDEMIC SUBSPECIES,
NOT THREATENED

Identification

The Chatham Island pipit is currently considered an endemic subspecies of the New Zealand pipit. However, recent research suggests that Chatham Island pipits, along with pipits from Antipodes, Auckland and Campbell Islands, may be a separate species, distinct from the mainland pipit. Further study is required. Pipits look similar to the introduced skylark, but they lack the skylark's small crest, and have a prominent white eyebrow. Pipits are more approachable than skylarks, and typically walk away from the observer, flicking their tails up and down.

Right: Chatham Island pipit, The Sisters.
Photo: Christopher Robertson (DOC).
Below: Chatham Island pipit carrying food for chicks, Rangatira.
Photo: Helen Gummer.

Distribution and ecology

Pipits are abundant on Chatham Island, Pitt Island, Rangatira, Mangere, Star Keys, The Forty Fours, The Sisters, and The Pyramid. They prefer open habitat, rough pasture and coastal areas, including rocky islands. There are no estimates of population size, but a study of densities on Chatham Island found that pipits were more numerous there than in similar North Island habitats.

Pipits are strongly territorial during the breeding season, with some breeding pairs remaining on territory all year. They eat a range of insects, spiders and sandhoppers, and also seeds of grasses, clover and other plants. Most food is taken from the ground, but some insects are caught on the wing. The female builds a deep cup nest, well concealed in vegetation. Two or three broods are typically raised between August and February.

Chatham Island pipit at nest,
Mangere, December 1981.
Photo: Rod Morris (DOC).

Threats and conservation

The Chatham Island pipit is not regarded as threatened. Despite the introduction of cats, three species of rats, mice and hedgehogs to Chatham Island, pipits have remained numerous. They have probably benefited from land clearance and the creation of more open country.

Skylark *Alauda arvensis* 18 cm

INTRODUCED TO NEW ZEALAND AND CHATHAMS

Skylarks were introduced to the main islands of New Zealand by European settlers during the late 1800s because of their melodious flight song. Some were introduced to Chatham Island by Walter Hood in the 1890s, but it is likely that many also flew in from mainland New Zealand. Between August and January, males give a loud trilling territorial song, which can last for up to five minutes, while they fly up and hover at 30-100 m. Skylarks are common in open country, sand dunes and farmland on Chatham and Pitt Islands and are also found on Rangatira and Mangere. Skylarks have very similar colouring to pipits but are slightly stouter, have a shorter tail and do not flick their tails in the way characteristic of pipits.

Skylark at nest.
Photo: Peter Morrison
(DOC).

Dunnock *Prunella modularis*   14 cm

INTRODUCED TO NEW ZEALAND

Also known as hedge sparrow, though not related to true sparrows. Both sexes are dainty, slim birds with dark brown backs and a grey wash on the head and breast. Hundreds of dunnocks were introduced to mainland New Zealand in the 1860–80s, and from there flew to many outlying island groups, arriving on the Chatham Islands before 1950.

Dunnock at nest.
Photo: Mike Soper (DOC).

Although inconspicuous, dunnocks are very abundant on the Chatham Islands, occurring in rough farmland, shrubland, and forests on Chatham and Pitt Islands, Rangatira, Mangere, and Tapuaenuku.

Blackbird *Turdus merula*   25 cm

INTRODUCED TO NEW ZEALAND AND CHATHAMS

Adult males are entirely black with a bright orange-yellow bill and eye-ring. Females are dark brown with a duller orange bill. The blackbirds' natural range includes Europe, northwest Africa, northern parts of Asia, and China. About 1000 birds were released in mainland New Zealand by Acclimatisation Societies. By 1900 they were widely distributed throughout the country, including the Chatham Islands, where they were introduced by Walter Hood. They are found on Chatham and Pitt Islands, Rangatira, Mangere, and Tapuaenuku, occurring in forests as well as farmland.

Right: Male blackbird at nest. Photo: Peter Reese. Far right: Adult female blackbird. Photo: Rod Morris (DOC).

Song thrush *Turdus philomelos*   23 cm

INTRODUCED TO NEW ZEALAND

Song thrush at nest. Photo:
Barry Harcourt (DOC).

Slightly smaller than a blackbird, the song thrush is brown above and
whitish below, with speckled dark brown spots on the breast. Song
thrushes originate from Europe and western and central Asia, and
were introduced to mainland New Zealand in the 1860s and 1870s,
arriving on the Chatham Islands about 1922. They are well established
throughout New Zealand and the Chatham Islands.

Yellowhammer *Emberiza citrinella*   16 cm

INTRODUCED TO NEW ZEALAND

Yellowhammers are sparrow-sized with reddish-brown upperparts and
yellow faces and bellies. Their natural range is from Britain to Siberia.
About 500 were introduced to mainland New Zealand in the 1870s

and 1880s, and they colonised
the Chatham Islands about 1910.
Yellowhammers inhabit open
country, but they are very rare
on Chatham and Pitt Islands, and
may not have a resident
population.

Male yellowhammer.
Photo: Rod Morris (DOC).

73

Chaffinch *Fringilla coelebs* 15 cm

INTRODUCED TO NEW ZEALAND

Chaffinches are sparrow-sized finches with brown, black, grey and (in the male) pinkish brown plumage. They have conspicuous white bars on the shoulder and the wing. Native to Europe, North Africa, the Middle East, and western and central Asia, chaffinches were introduced to mainland New Zealand by Acclimatisation Societies during the late 1800s, and colonised the Chatham Islands before 1950. They are widespread in New Zealand, including Chatham and Pitt Islands, Rangatira, Mangere, and Tapuaenuku.

Above: Female chaffinch. Photo: Mike Soper (DOC). Right: Male chaffinch at nest. Photo: T. Smith (DOC).

Greenfinch *Carduelis chloris* 15 cm

INTRODUCED TO NEW ZEALAND

Greenfinches are olive-green sparrow-sized birds that are native to Europe, North Africa, the Middle East and western Asia. They were introduced to mainland New Zealand during the 1860s and are now abundant throughout New Zealand. Greenfinches are locally common on the Chatham Islands, which they colonised before 1920.

Male greenfinch. Photo: John Kendrick (DOC).

74

Goldfinch *Carduelis carduelis* 13 cm

INTRODUCED TO NEW ZEALAND

Goldfinches are small brown, black and white birds with gold bars on their wings, and bright red, white and black faces. Their natural range is Europe, North Africa, the Middle East and western Asia. About 500 goldfinches were liberated in mainland New Zealand in the late 1800s, from where they colonised the Chatham Islands about 1910. They are common on Chatham and Pitt Islands, and straggle to other islands in the group.

Goldfinch at nest.
Photo: (DOC).

Redpoll *Carduelis flammea* 12 cm

INTRODUCED TO NEW ZEALAND

Redpolls are small mottled-brown finches with a crimson patch on the forehead. In the breeding season the male also has a pink flush on the breast. From North America, Europe and Asia, redpolls were introduced to mainland New Zealand in the late 1800s. They established throughout the mainland, and are now by far the most common finch species on Chatham and Pitt Islands, Rangatira, Mangere, Tapu-aenuku, and Star Keys.

Redpoll at nest. Photo: Peter Reese.

House sparrow *Passer domesticus* 14 cm

INTRODUCED TO NEW ZEALAND

House sparrows are a common associate of humans throughout much of the world. Males have a chestnut-and-black back with grey-white underparts and a black bib. Females are a lighter brown. House sparrows were introduced to New Zealand, and colonised the Chatham Islands about 1880. They are found on Chatham and Pitt Islands, and occasionally Rangatira, usually close to dwellings.

Adult male house sparrow.
Photo: Jeremy Rolfe.

Starling *Sturnus vulgaris* 21 cm

INTRODUCED TO NEW ZEALAND AND CHATHAMS

Starlings have glossy black plumage, speckled with buff and white spots. They have a short tail and pointed wings. They were introduced to New Zealand in the late 1800s, and to Chatham Island by Walter Hood before 1900. Starlings are very numerous in the Chatham Islands, and breed on most islands in the group. Large flocks are seen at Kaingaroa, and flying between Pitt Island and Rangatira.

Adult male starling.
Photo: Dick Veitch (DOC).

76

Forest birds

Forest interior, Chatham Island. Photo: Richard Suggate (DOC).

The endemic forest birds of the Chatham Islands have suffered greatly from habitat clearance and the introduction of cats and rodents. The only species readily seen in most forest remnants on Chatham Island is the fantail, which also visits gardens. South of the Waitangi–Owenga road, the more extensive forests also contain parea, Chatham Island red-crowned parakeets, and Chatham Island warblers. All the sites are difficult to access, and require landowner permission. However, all three species (and especially parea) may be seen near the public road end at Awatotara. The parakeet and warbler are easily seen on Pitt Island, along with tui and tomtit. A predator-fenced site there is the focus for attempts to re-introduce black robin. In time, snipe and Forbes' parakeets may also be re-introduced to sites where they can be seen and appreciated by Chatham Islanders and their guests.

Chatham Island snipe
Coenocorypha pusilla

20 cm

CHATHAM ISLANDS ENDEMIC, RANGE RESTRICTED

Chatham Island snipe,
Rangatira.
Photo: Colin Miskelly.

Identification

The Chatham Island snipe is a small squat bird with a long bill, and rich dappled brown, rust and black plumage. The lower breast and belly are pale and unmarked. Snipe are usually seen as singles or pairs, quietly probing for invertebrates on the forest floor. When startled, they erupt into flight with whirring wings, but generally do not fly far.

Distribution and ecology

The different forms of New Zealand snipe were formerly distributed throughout New Zealand and the Chatham Islands, but are now confined to remote islands free of introduced mammal predators. Chatham Island snipe are currently restricted to Rangatira, Mangere, Tapuaenuku, and Star Keys, but are occasionally seen on Pitt Island. The total population is estimated to be about 1000 pairs. Snipe inhabit a wide range of vegetation types, including forest, scrub and tussock. They prefer sites with moist soils and dense ground cover, as all food is obtained by probing. The two well-camouflaged eggs are laid in September–March in a nest constructed among dense vegetation on or near the ground. Chicks leave the nest on the day of hatching; the brood is split between the parents, who each feed one chick until it becomes independent.

Chatham Island snipe at nest, Mangere, October 1980. Photo: Rod Morris (DOC).

Threats and conservation

Chatham Island snipe were exterminated on Mangere by feral cats in the 1890s, but the cats themselves died out in the 1950s (largely as a result of shooting by visiting sheep shearers), and 23 snipe were successfully re-introduced from Rangatira in 1970. From Mangere, snipe recolonised Tapuaenuku. Kiore are presumed to have eradicated snipe from mainland New Zealand and many offshore islands, including Chatham Island. Rat invasion of their remaining island habitats is the greatest ongoing threat to snipe populations in the Chatham Islands.

The protection of Rangatira and Mangere as Nature Reserves, the removal of stock, and the implementation of quarantine measures to prevent the arrival of mammalian predators to the islands, have all benefited snipe. In the 1980s there were two attempts to establish Chatham Island snipe in captivity at the National Wildlife Centre, Mount Bruce. Eggs and live birds were both transferred. However,

Chatham Island snipe chick, Rangatira, February 2004. Photo: Don Merton.

problems were experienced getting the birds to eat artificial food, and none survived to breed.

In the future it is intended that snipe be established on Pitt Island, inside the predator-proof fence built at Ellen Elizabeth Preece Conservation Covenant. In the long term other predator-free sites may be created in the Chatham Islands, allowing snipe to be returned to more of their historic range. Chatham Island snipe may also be used to replace the extinct North Island snipe at predator-free sites.

Juvenile Chatham Island snipe, Rangatira, January 1984. Photo: Colin Miskelly.

Parea
Hemiphaga chathamensis 55 cm

CHATHAM ISLANDS ENDEMIC, NATIONALLY CRITICAL

Other names: Chatham Island pigeon, pigeon

Parea on nest, Chatham Island, October 1993. Photo: Ralph Powlesland (DOC).

Identification
The parea is one of the world's heaviest pigeons and is about one-fifth heavier than the New Zealand pigeon (kereru/kukupa; *H. novaeseelandiae*). Its dorsal plumage and upper breast is more purple and pearl-grey than the New Zealand pigeon, but it has the same white lower breast, shoulder straps and belly. The bill is red with an orange tip (the latter is lacking in New Zealand pigeon). Sexes are alike. Parea fly with noisy wing-beats, and during the breeding season they perform conspicuous display dives, flapping upwards from their perch, then stalling and diving sharply down. The parea was until recently considered a subspecies of the New Zealand pigeon. Based on its larger size, plumage differences, and its confinement to the Chatham Islands, the parea is now given full species status.

Distribution and ecology
Parea were formerly widespread and common on Rangatira, Mangere, Chatham and Pitt Islands. By 1938, few were seen in northern Chatham Island, but they were moderately plentiful in the more extensive areas

of forest in the south. Parea disappeared from northern Chatham Island forests in the 1970s, although there have been a number of sightings there in recent years. They disappeared from Mangere and Rangatira over 100 years ago when the islands were largely cleared for farming, and the Pitt Island population apparently crashed following a large forest fire in the mid 1900s. Currently, only a few parea are thought to be present on Pitt Island, and there has been no recorded breeding for many years. Two birds are present on Rangatira from a transfer there in the 1980s.

Parea have an important role in maintaining healthy forest structure: many Chatham Island tree species are thought to be dependent on parea for seed dispersal. Parea can breed all year round, but nest predominately during winter and spring, laying a single white egg in a robust nest of twigs 0–10 metres off the ground. The timing of nesting and the proportion of pairs that breed vary between years in response to the abundance and quality of available food, particularly hoho fruit.

Threats and conservation

The decline of parea had a number of causes, including loss of forest habitat, predation of adults and chicks by feral cats, and probably predation of eggs and chicks by rodents and possums. Competition from browsing animals, particularly possums, is another factor in their decline, and hunting by people for food may have also contributed. With the exception of hunting, these causes of past decline still threaten the population today.

Parea, Chatham Island.
Photo: John Mason.

Concern about the decline in parea numbers was raised during the 1970s. In 1983, 13 parea were translocated from southern Chatham Island to Rangatira, but only a couple of birds remained on the island, and successful breeding was not recorded. The parea population reached its lowest level of about 40 birds in 1990, confined to forest in southern Chatham Island. A 4-year research programme, habitat protection and predator control led to an increase to about 150 birds by 1995. Habitat protection and predator control are ongoing in the Tuku Nature Reserve and adjacent conservation covenants to protect both parea and taiko. There is no precise estimate of the total population today, but the number of sightings of parea in southern Chatham Island indicates that the population has continued to increase slowly. A survey in 1999 found a slight increase in numbers in areas where the forest was protected and there was predator control; there was a decline in areas further south that were not receiving protection.

Forbes' parakeet
Cyanoramphus forbesi

28 cm

CHATHAM ISLANDS ENDEMIC, NATIONALLY ENDANGERED

Forbes' parakeet, Mangere.
Photo: Dave Crouchley
(DOC).

Identification

Forbes' parakeet is a medium-sized, bright-green parakeet with a long tail and orange-red eyes. It has a yellow crown to just behind the eye, and a narrow red frontal band above the bill. The wing coverts and outer primaries are violet-blue. Sexes are alike, although the female is slightly smaller. The taxonomic status of Forbes' parakeet has been contentious. Originally described as a distinct species, for many years it was regarded as a subspecies of the yellow-crowned parakeet (*C. auriceps*), and was sometimes referred to as the Chatham Island yellow-crowned parakeet. Recent genetic work has found that Forbes' parakeet is highly differentiated from all other New Zealand parakeets, and should be considered a full species.

Distribution and ecology

Forbes' parakeet is endemic to the Chatham Islands and is one of the rarest of New Zealand's parakeets. Historically, Forbes' parakeets were found on Mangere and Tapuaenuku, and also visited Pitt Island seasonally to feed. However, by 1930 the entire population was confined to Tapuaenuku. Mangere was recolonised following the disappearance of feral cats and the regeneration of native vegetation after the removal of grazing animals in 1968. Forbes' parakeets are now confined to Mangere and Tapuaenuku, with vagrants reported from southern Chatham Island, Rangatira, and Pitt Island.

On Mangere, Forbes' parakeets occur at highest densities in the small patch of remnant forest. However, forest habitat is extremely scarce on Mangere, and parakeets also use open habitats such as scrub and grassland. Nesting occurs in November–January; most nests are in cavities in Chatham Island akeake trees. The male feeds the female while she incubates the 2–9 eggs, and both parents feed the chicks.

Threats and conservation

Destruction of forest habitat and predation by feral cats and rats were probably the key causes of the decline of Forbes' parakeet. Another major threat to the Forbes' parakeet population was hybridisation with Chatham Island red-crowned parakeets. After cats and stock were removed from Mangere, the red-crowned parakeet population expanded rapidly at the same time as Forbes' parakeets were recolonising from Tapuaenuku. Loss of forest may have given the red-crowned parakeet a competitive advantage, as they make more use of open habitats. Hybridisation between the two species led to genetic swamping of Forbes' parakeet by the more numerous red-crowned parakeets. Forbes' parakeets continue to be limited by habitat availability, and hybridisation with red-crowned parakeets. Further habitat loss from accidental fire, and accidental introduction of mammalian predators such as rodents or cats to Mangere, are ongoing threats.

Forbes' parakeet, Mangere.
Photo: Dick Veitch (DOC).

Efforts to save Forbes' parakeets from complete hybridisation with red-crowned parakeets began in 1976 with the start of a revegetation project on Mangere. The removal of both hybrid and red-crowned parakeets from Mangere was also undertaken, as this was considered essential to ensure the survival of a distinct Forbes' parakeet. Revegetation has continued to the present day at a rate of approximately 6000 trees per annum. Intermittent culls of red-crowned and hybrid parakeets continued until 1999, when a moratorium was put in place while ecological and genetic research on Mangere parakeet populations was undertaken. This research will be used to guide management to ensure the survival of Forbes' parakeet with minimal intervention. A population of around 900 parakeets was estimated to be present on Mangere in 2002, with more than 83 percent being "Forbes-like" in appearance. Options for re-introduction to other sites are limited, as red-crowned parakeets are common on Pitt Island and Rangatira, and are present in southern Chatham Island. Forest remnants lacking parakeets are present in northern Chatham Island, but would require predator fencing before Forbes' parakeets could be re-introduced.

Forbes' parakeet, Mangere.
Photo: Dave Crouchley (DOC).

Chatham Island red-crowned parakeet
Cyanoramphus novaezelandiae chathamensis

28 cm

CHATHAM ISLANDS ENDEMIC SUBSPECIES,
RANGE RESTRICTED

Other name: Chatham Island red-crowned kakariki

Chatham Island red-crowned
parakeet, Rangatira. Photo:
Dave Crouchley (DOC).

Identification

The Chatham Island red-crowned parakeet has bright-green plumage, with a crimson forehead and crown to above the eye, and a small red patch just behind the eye. There is a red patch on the side of the rump; the wing coverts and some of the outer flight feathers are edged with violet-blue. Sexes are alike, although the male is slightly larger with a heavier bill. The Chatham Island red-crowned parakeet is slightly larger than the New Zealand red-crowned parakeet (*C.n. novaezelandiae*). Parakeets are distinctive in flight, with their direct, rapid flight over the canopy or through trees, and their long tails. Their call is a rapid loud chatter.

Distribution and ecology

Chatham Island red-crowned parakeets are endemic to the Chatham Islands. They are currently found on Chatham and Pitt Islands and Rangatira, and in small numbers on Mangere and Tapuaenuku. Red-crowned parakeets declined to very low numbers on Chatham Island by the late 1960s, and were mainly confined to the southern forests. Numbers appear to have

Chatham Island red-crowned parakeet at nest hole.
Photo: Dick Veitch (DOC).

recovered in southwestern Chatham Island in response to ongoing possum and feral cat control. Red-crowned parakeets have been observed occasionally in fenced forest remnants in northern Chatham Island. They are also common on Pitt Island; however, their stronghold is on Rangatira which supports a large population.

Red-crowned parakeets use open habitats such as forest margins, scrub and grassland as well as forest. They feed on leaves, shoots, flowers and seeds, but also take invertebrates, nectar and fruit. Nests are built in holes in trees, or occasionally in rock crevices or under dense vegetation. Breeding ecology is similar to that of Forbes' parakeet.

Threats and conservation

Ongoing deterioration of Chatham and Pitt Island forests caused by browsing mammals, predation by feral cats on Chatham and Pitt Island, and rat predation on Chatham Island continue to limit red-crowned parakeet population recovery. The population on Mangere was culled during the 1970s–90s to protect the nationally endangered Forbes' parakeet by reducing hybridisation. This programme was stopped in 1999 while ecological and genetic research was undertaken on the Mangere parakeets.

Chatham Island red-crowned parakeet, Rangatira.
Photo: Colin Miskelly.

The legal protection of Rangatira, and measures implemented to ensure the island remains free of introduced pests, provide important protection for red-crowned parakeets in their stronghold. Protection of forest habitat on Chatham and Pitt Islands by fencing to exclude stock, and the control of possums (Chatham Island only) and feral cats will allow expansion of red-crowned parakeets into areas they formerly occupied.

Shining cuckoo *Chrysococcyx lucidus* 16 cm

NEW ZEALAND NATIVE, NOT THREATENED

Known as pipiwharauroa by Maori, the arrival of the shining cuckoo from the Pacific in September heralds the spring. This small but striking bird, with metallic green upperparts with bronze tones, and barred white and green underparts, is more often heard than seen. Shining cuckoos come to New Zealand to breed, in usual cuckoo fashion, by laying their eggs in other birds' nests to be reared by the host. They are uncommon in the Chatham Islands, where they lay in nests of the Chatham Island warbler.

Shining cuckoo. Photo: John Kendrick (DOC).

Silvereye *Zosterops lateralis* 12 cm

NEW ZEALAND NATIVE, NOT THREATENED

Named for the distinctive white ring around its eye, this small, predominantly olive-green bird is also called white-eye or waxeye. Silvereyes arrived in New Zealand from Australia in the mid 1800s, hence their Maori name of tauhou (stranger). Abundant on Chatham and Pitt Islands, Rangatira, and Mangere, loose flocks are often seen in winter, feeding on fruit, nectar and insects in gardens and forest habitats.

Silvereye. Photo: John Kendrick (DOC).

Chatham Island warbler
Gerygone albofrontata 12 cm

CHATHAM ISLANDS ENDEMIC, RANGE RESTRICTED

Identification

The Chatham Island warbler is a small forest bird, with olive-brown upperparts, white sides and underparts and a short, fine bill. The male has a prominent white face, including forehead and throat, with a contrasting dark eye-stripe; the flanks and undertail are also whitish. Females are smaller with duller plumage; the face, underparts and sides being grey-white with a tinge of yellow, particularly on the throat. Only the male gives a full song. Warblers spend most of their time foraging under the canopy, gleaning invertebrates from leaves and tree crevices.

Chatham Island warbler
sunbathing, Rangatira.
Photo: Helen Gummer.

Chatham Island warbler in nest, Mangere, December 1982. Photo: Dave Crouchley (DOC).

Distribution and ecology

The Chatham Island warbler is common in forests on southern Chatham Island, Pitt Island, Rangatira, Mangere, Tapuaenuku, and Star Keys, and also occurs on The Castle, and possibly the inner Murumurus. There is no estimate of total numbers. Warblers have not spread into modified habitats, but inhabit the full range of native forest types. Warblers are absent from the small forest remnants in northern Chatham Island. They occur at high densities on Rangatira and Mangere compared to southern Chatham Island. Nests are distinctive enclosed domes, with a small entrance on the side, generally suspended in trees or tall shrubs. The 2–4 eggs are laid in September–December and are incubated by the female, but both parents feed the chicks. A single brood is raised each year.

Threats and conservation

Loss of forest habitat through burning and clearance for farmland, and habitat deterioration due to introduced browsers (particularly the loss of forest understorey), are likely to have caused the decline of warblers through much of their range. The fragmentation and degradation of forest habitat in north Chatham Island is likely to be the reason warblers are no longer found there. The presence of predators may explain the lower densities of warbler on southern Chatham Island compared to Rangatira and Mangere. Like the grey warbler (*G. igata*) in New Zealand, Chatham Island warblers occasionally act as unwitting foster parents for shining cuckoos.

Protection of forest habitats throughout the Chatham Islands will assist warbler population recovery, although the ability of warblers to re-colonise isolated forest remnants unaided is uncertain. The maintenance of habitat free from browser pressure, allowing development of a dense understorey and deep leaf litter, will be essential for populations to establish and survive. During the early 1980s, warblers were trialled as foster parents during the black robin recovery programme on Mangere, but were found to be unsuitable.

Male Chatham Island warbler. Photo: Dave Crouchley (DOC).

Chatham Island fantail
Rhipidura fuliginosa penita   16 cm

CHATHAM ISLANDS ENDEMIC SUBSPECIES,
NOT THREATENED

Chatham Island fantail,
Chatham Island.
Photo: Peter Reese.

Identification
The Chatham Island fantail is a small flycatcher with a rounded head, short bill and long tail which is often fanned. Sexes are alike. It is a subspecies of the New Zealand fantail, and looks similar to pied fantails (*R. f. fuliginosa* and *R. f. placabilis*) of mainland New Zealand, but with more white in the tail.

Distribution and ecology
Fantails are common in most areas of native forest, scrub and gardens on Chatham and Pitt Islands, and occasionally breed on Rangatira and Mangere, straying to Star Keys. While fantails can be abundant at times, there are frequent population fluctuations, with sharp declines during severe weather. They eat insects, and forage with distinctive flitting movements, their tail fanning and closing as they manoeuvre through forest and gardens. Compact, open-topped nests are built 1.5–5 metres above the ground, generally in the understorey, in October–January. Both sexes incubate the 3–4 eggs and brood the chicks.

Threats and conservation
The deterioration and loss of forest habitat through burning and clearance for farming, and the introduction of browsers are likely to have been key factors in the decline of fantails. Recently observed declines and recoveries of fantail populations appear to be in response to climatic conditions, where storms lasting many days make it difficult for fantails to find sufficient prey. Protection of forest habitats will assist maintenance of fantail populations on Chatham Island. Reduced browser pressure allows development of a dense, sheltered understorey and deep leaf litter, which encourage healthy insect populations considered essential for fantails to survive long-term.

Chatham Island fantail at nest, Chatham Island, October 1993.
Photo: Peter Reese.

Chatham Island tomtit
Petroica macrocephala chathamensis

13 cm

CHATHAM ISLANDS ENDEMIC SUBSPECIES,
NATIONALLY ENDANGERED

Other name: Chatham Island tit

Male Chatham Island tomtit,
Rangatira.
Photo: Colin Miskelly.

Female Chatham Island
tomtit, Rangatira.
Photo: Colin Miskelly.

Identification
The Chatham Island tomtit is a subspecies of the New Zealand tomtit. It is a small forest bird with a large head and short tail. It is most similar to the South Island tomtit (*P. m. macrocephala*), though slightly larger. The male has a glossy black head with a small white spot above the bill, and black upperparts and upper breast. The underparts are yellowish, getting brighter and more orange towards a sharp dividing line between the yellow and black in the middle of the breast. There is a white wing-bar and sides to the tail. The female has a brown head and upperparts, with a white frontal spot above the bill. The throat, breast and belly are cream, and the wing-bar and sides of the tail are white.

Distribution and ecology
Tomtits are currently restricted to Pitt Island, Rangatira, and Mangere, with a total population of about 1000 birds. On Chatham Island, tomtits declined to low numbers in the southern part of the island as early as 1938, and had disappeared by the mid 1970s. They inhabit mature and regenerating native forest and scrub, where they feed on small

invertebrates mostly in the mid and upper tiers of the forest. Nests are in tangles of pohuehue vines or in hollow branches or tree cavities, 0.5–8 metres above the ground. Males feed the females while the latter incubate the 2–4 eggs, which are laid in October–December.

Threats and conservation

The deterioration and loss of forest habitat through burning and clearance for farming, and the introduction of browsers are likely to have been key factors in the decline of tomtits. Predation by feral cats and rodents, and possibly possums preying on eggs and chicks, are likely to have contributed to their disappearance on Chatham Island. Predation is thought to be the main factor limiting tomtit population recovery on Chatham and Pitt Islands today.

Chatham Island tomtits are best known for their role as foster parents of black robin eggs and chicks during the 1980s. As part of this programme, all tomtits were removed from Mangere in 1976 to reduce competition with the tiny black robin population. Tomtits were successfully re-introduced to Mangere in 1987–89. Tomtits were also re-introduced to the Tuku Valley, Chatham Island, in 1998, in the hope that the predator control conducted for taiko and parea would have improved the habitat for other forest birds. However, none is known to have survived, suggesting that rat predation may be an important factor preventing the re-establishment of tomtits on Chatham Island. A predator-proof fence constructed on Pitt Island has provided an area of cat-free habitat in which tomtits have thrived.

Black robin *Petroica traversi* 15 cm

CHATHAM ISLANDS ENDEMIC, NATIONALLY CRITICAL

Other names: kakaruai, Chatham Island black robin

Black robin, Rangatira.
Photo: Colin Miskelly.

Identification

The black robin is a small songbird with completely black plumage, becoming blackish-brown as the feathers wear. It has a short fine, black bill, long thin dark legs and an upright stance. Sexes are alike although the female is slightly smaller. Black robins feed mostly on the forest floor. Their call is a high-pitched single note, and the male song is a clear simple phrase of 5–7 notes. It most closely resembles the Snares Island black tomtit (*P. macrocephala dannefaerdi*), but its behaviour shows its relationship to the New Zealand robin (*P. australis*).

Distribution and ecology

Black robins were originally present on Mangere, Tapuaenuku, Chatham and Pitt Islands. There are no historical records of the species' presence on Rangatira, although it is likely that they did occur there. By 1872, when the species was first encountered by European observers, it had already disappeared from Chatham Island. For many years the world population of black robins was confined to Tapuaenuku, a tiny Maori-owned island, protected by extremely tall cliffs.

93

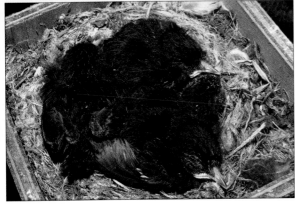

Black robins are insectivorous forest-dwellers, although they also make use of lower scrubby vegetation. They are currently confined to Rangatira and Mangere, with a total population of approximately 200 adult birds. A few have recently been reintroduced to Pitt Island. On Mangere, the robins are confined to an 8 ha forest remnant, where they are at high population densities. On Rangatira, they inhabit all forested areas, with more birds in the lower, north-facing forest than the more exposed, predominately ribbonwood forest of the southern part of the island. Territories are maintained year-round, and the same mate is usually kept year after year. Nests are a neat cup lined with moss

Top: Black robin nest with three eggs, Rangatira, January 2000. Above: Fourteen-day-old black robin chicks in a nest box, Rangatira, January 1987. Photos: Don Merton (DOC).

and feathers, usually in a tree cavity, hollow branch or rotten stump. The 1–3 eggs are laid in October–November, and typically a single brood is reared. The female does all the incubating, although she is fed by the male, and both parents feed the chicks.

Threats and conservation

Mammalian predators, particularly rodents and feral cats, are likely to have been the main cause of the black robin's decline. Cats exterminated the black robin on Mangere by about 1900. Habitat loss was also likely to have been important. Large areas of forest habitat on the main islands were destroyed by fire and conversion to farmland before the end of the 1800s. Introduced browsers accelerated the degradation of bush remnants.

Current threats to black robin on Rangatira and Mangere include the risk of invasion by introduced predators or disease, competition from avian competitors such as starlings, and habitat loss through fire. Hybridisation between black robins and Chatham Island tomtits was

Juvenile black robin, Rangatira.
Photo: Don Merton.

recorded on a number of occasions, but in all instances involved black robins that were reared by tomtits; none of these foster-reared birds or hybrids remains in the population. The rescue of the black robin from its last tiny refuge on Tapuaenuku and the subsequent recovery of the population is now a celebrated episode in New Zealand conservation history. Efforts to save the black robin from extinction began in the late 1970s, when the tiny population (down to seven birds) was moved from Tapuaenuku, where the small forest patch they inhabited was deteriorating, to Mangere. On Mangere, an intensive programme of manipulating black robin breeding attempts to boost productivity was initiated. This involved using firstly warblers and then, on Rangatira, tomtits as foster parents, so that the robins could be induced to produce multiple clutches of eggs each season. The success of this programme, and particularly the cooperation of the then only remaining productive female 'Old Blue', meant that by 1983 a second population could be started on Rangatira.

When black robin numbers reached 80 in 1989, the decision was made to discontinue the intensive nest manipulation programme, to see if the population would increase on its own. Numbers continued to increase, and by 1998 it was thought large enough (over 200 birds) for a less intensive monitoring programme to be initiated.

The recovery programme for black robins aims to establish a third population in the Chatham Islands. To this end, a predator-proof fence has been constructed around 40 ha of regenerating bush on Pitt Island. A first transfer of 14 birds to this site took place in September 2002. Unfortunately none of these birds established at the site. A further 20 juveniles were released in March 2004.

On Rangatira and Mangere, strict quarantine measures are applied to protect black robins from exposure to mammalian predators, and there is an ongoing planting programme on Mangere to increase the area of forest available for the robins. With further regeneration, both Rangatira and Mangere will be able to support greater numbers of robins than they do currently.

Chatham Island tui
Prosthemadera novaeseelandiae chathamensis

32 cm

CHATHAM ISLANDS ENDEMIC SUBSPECIES,
NATIONALLY ENDANGERED

Other names: koko, parson bird

Identification

Chatham Island tui have black plumage with green, bluish-purple and bronze iridescent sheens when seen in sunlight. Two tufts of white feathers curl at the throat. Sexes are alike. The Chatham Island tui is a subspecies of the tui found on mainland New Zealand; it is a larger bird with longer throat tufts. The song is quite different from the New Zealand tui, with more guttural coughing and whistles. Tui flight is energetic, with acrobatic dives and noisy, whirring wing beats.

When feeding on flax nectar, tui often have an orange dusting of flax pollen on their foreheads.

Distribution and ecology

Tui were formerly widespread on Chatham and Pitt Islands, Rangatira, and Mangere, and possibly Tapuaenuku. By 1938, tui were recorded as being less common on northern Chatham Island, common in southern areas, Pitt Island and Rangatira, but absent from Mangere (which had recently been cleared). By the 1970s, tui on Chatham Island were restricted to low numbers

Chatham Island tui on flowering Chatham Island flax, Rangatira. Photo: Helen Gummer (DOC).

in the south, and by the early 1990s they had disappeared apart from the occasional bird crossing Pitt Strait in late summer and winter. Tui are now confined to Rangatira and Pitt Island, with a few birds present on Mangere.

The tui population on Rangatira is estimated to be around 250 birds. The size of the Pitt Island tui population is unknown. Breeding is known to occur on both Rangatira and Pitt Islands, but most breeding appears to take place on Rangatira. Many tui leave Rangatira during winter and move to Pitt Island, returning to breed on Rangatira the following spring. Tui feed on nectar, fruit and insects. They are important for the pollination and seed dispersal of native forest plants. The tui is the only remaining honeyeater in the Chatham Islands, following extinction of the Chatham Island bellbird in the early 1900s. Details of breeding are poorly known, but apparently more birds breed in good flax flowering years. On Rangatira, most nest in thick tangles of pohuehue vines over forest trees. The breeding season is November–January; 2–4 eggs are laid, and typically 1–3 fledglings are reared.

97

Chatham Island tui on flowering Chatham Island flax, Rangatira. Photo: Colin Miskelly.

Threats and conservation

Loss of habitat, and predation by feral cats, rodents and possums are likely to have been the main causes of decline of tui. While there are significant areas of forest habitat remaining on southern Chatham Island, the combined pressure of predators and browsers appears to have been too much for tui to survive. The disappearance of tui from Chatham Island coincided with the spread of possums, and the associated loss or severe reduction of plant species that provide flowers and fruit important to tui.

The legal protection of Rangatira as a Nature Reserve, and measures to protect the island from the introduction of pests, has provided secure habitat for tui. Research conducted on Rangatira in 1995–98 provided information on tui ecology and breeding. The protection of forest habitat and reduction of predators and browsers on Chatham Island, and the revegetation programme on Mangere, should provide suitable habitat for the re-establishment of tui in the future.

Glossary

Biennial   Every two years

Brood (noun)   All the chicks hatched from a single clutch of eggs

Brood (verb)   To cover chicks to keep them warm

Browser   An animal (usually mammal) that eats leaves or twigs of trees and shrubs

By-catch   Non-target captures. In this context, seabirds caught on hooks set for fish

Carpal joint   The outermost joint/bend in the wing, equivalent to our wrist

Circumpolar   Present or moving all around the world in a high latitude (near the pole)

Clutch   All the eggs in one nest, or laid by a single female during a single breeding attempt

Coloniser   Threat category used by the Department of Conservation for taxa that have arrived in New Zealand without help from humans and have been successfully reproducing in the wild for less than 50 years

Conservation Covenant   A formal agreement to protect freehold land for conservation purposes, registered on the title, where the land remains in private ownership. Survey and fencing costs met by the Nature Heritage Fund

Coverts   Small feathers that overlay the bases of the main flight feathers, to smooth the flow of air over wing and tail surfaces

Dorsal   On or referring to the back or upper surface

Endemic   Naturally breeding only at that location

Feral   Wild individuals or populations of animal species that are usually farmed or kept in domestication, e.g. cats, cattle, sheep and pigs on the Chatham Islands

Fledging   Leaving the nest, or becoming capable of flight

Fledgling   A young bird that has recently learnt to fly (for most species, this coincides with leaving the nest)

Frontal shield   Horny extension of the beak onto the forehead, typically found in gallinules (e.g. pukeko) and coots

Gadfly petrels   Seabirds of the genus *Pterodroma* ("wing runners") renown for their fast, impetuous flight, reminiscent of a gadfly (horse-fly)

Gazettal   In this context, the part of the formal process announcing the status of a parcel of Crown Land, i.e. land owned by the people of New Zealand, and managed under the Reserves Act, by publishing in the New Zealand Gazette

Genus   First word of a scientific name, used to group closely related species. For example, black robin and tomtit both belong to the genus *Petroica*

Gradual Decline   Threat category used by the Department of Conservation for taxa that have large populations and wide distributions, but are believed to be declining gradually at most or all sites

Guano   Bird (or bat) faecal matter, especially where this accumulates in large deposits

Hopo   Moriori term for chicks of northern royal albatross and other albatross/ mollymawk species

Insectivore/insectivorous   An animal that eats insects and other small terrestrial invertebrates

Invertebrate   An animal that does not have a backbone (vertebral column)

Iwi   Maori and Moriori tribe or tribes

Kawenata   Maori for Covenant. Term used for conservation covenants established on land owned by Maori/Moriori, with survey and fencing costs met by Nga Whenua Rahui fund

Locally extinct   A species that has died out at the location referred to, but still survives elsewhere

Migrant   Threat category used by the Department of Conservation for animals that predictably and cyclically visit New Zealand as part of their normal life cycle, but do not breed here

Monogamous   Typical breeding system of birds (and people), where each bird has a single mate of the opposite sex at any one time

Moriori   The original people of the Chatham Islands. Many Chatham Islanders are descended from, and identify themselves as, Moriori

Muttonbirds   Large chicks of burrow-nesting petrels taken for human consumption, these days mainly used to identify sooty shearwaters, both as a species and for the harvested chicks

Nasal tubes   External nostril tubes found in all albatrosses and petrels, including mollymawks, shearwaters, fulmars, prions, and taiko

Nationally Critical Highest threat category used by the Department of Conservation; applies to species under imminent threat of extinction

Nationally Endangered Second highest threat category used by the Department of Conservation; applies to species likely to become extinct without active management

Nationally Vulnerable Third highest threat category used by the Department of Conservation; applies to species likely to become Nationally Endangered without active management

Native Naturally occurring (without human intervention)

Ngati Mutunga A sub-tribe of the Taranaki iwi Te Ati Awa that invaded the Chatham Islands in 1835, along with Ngati Tama. Only Ngati Mutunga chose to remain; many Chatham Islanders identify themselves as Ngati Mutunga

Nocturnal Active or occurring only at night

Not Threatened Threat category used by the Department of Conservation for taxa that have large populations and wide distributions, and are believed to be stable or increasing in numbers

Passage migrant A species likely to occur only briefly at the location referred to, when in transit between distant breeding and non-breeding grounds

Petrel Birds belonging to the families Procellariidae (shearwaters, gadfly petrels, prions, fulmars, diving petrels) and Hydrobatidae (storm petrels); 13 species breed on the Chatham Islands. Sometimes used for the entire order Procellariiformes, also including the family Diomedeidae (albatrosses and mollymawks), of which 7 species breed on the Chatham Islands

Primaries Large, outermost flight feathers of the wing

Prospecting In this context, searching for nest sites

Quarantine In this context, measures used to prevent rodents and other pests establishing on islands. This may include the practice of packing all food and equipment inside a mouse-proof store on the mainland, and unpacking it inside a mouse-proof room on the island

Rakau momori Traditional Moriori carvings on living kopi trees

Range Restricted Threat category currently used by the Department of Conservation for species that may be abundant, but naturally have a small range (typically an island or island group), and are therefore at risk of extinction if a new threat reaches the site

Raptor   A bird of prey (e.g. eagle, hawk, falcon)

Raukura   Albatross feather emblem used by both Moriori followers of Nunuku, and Te Ati Awa (Maori) supporters of Te Whiti O Rongomai

Serious Decline   Threat category used by the Department of Conservation for taxa that have large populations, but are declining rapidly

Sparse   Threat category used by the Department of Conservation for species that have a wide distribution, but are nowhere common

Species   The basic unit of animal (and plant) classification; a group of similar individuals or populations sufficiently closely related to be potentially capable of interbreeding, and recognised as distinct from all other groupings. Species are denoted by a binomial ("two names") scientific name, e.g. *Petroica traversi* for black robin

Speculum   A patch of brightly coloured upper-wing coverts with metallic reflections on the inner wing of some ducks (blue in mallard, green in grey duck)

Subfossil   Bones of birds that have been dead for 100s if not 1000s of years, but have not yet become fossilised. Such remains occur abundantly in some dune and cave deposits on the Chatham Islands

Subspecies   A subdivision of a species, usually considered capable of breeding with other subspecies of the same species, but typically isolated geographically. Subspecies are denoted by a trinomial ("three names") scientific name, e.g. *Prosthemadera novaeseelandiae chathamensis* for Chatham Island tui

Taonga   Valued item, treasure

Tapu   Sacred, or referring to ceremony to invoke spiritual security

Taxa (plural)/**taxon** (singular)   A unit of any rank in a classification system. Here we use "taxa" as a collective term referring to either or both species and subspecies

Taxonomy   The science of formally describing species/ subspecies and giving them a scientific name and classification. Hence **taxonomic**

Tchakat henu   Moriori term for themselves as indigenous people (equivalent to Maori *tangata whenua* – people of the land)

Telemetry   Determining the location of an animal by attaching a radio transmitter and then obtaining directional fixes using two or more radio receivers or receiver locations

Temperate   Those latitudes lying between the subtropics and the subantarctic (in the Southern Hemisphere) or subarctic (Northern Hemisphere)

Territorial   Describing an animal that defends its home range against others of its own species (and sometimes other species)

Translocation   Deliberate transfer of a species to a new site, these days usually done to improve conservation status

Type specimen   The original specimen used when a species is described and given a scientific name. The name remains associated with that one specimen.

Vagrant   An individual naturally occurring beyond its usual range

Waka korari   Moriori wash-through canoe-raft

Wattle   Fleshy, usually brightly coloured, growths from (or on) the faces of some birds

Appendix 1

CHECKLIST OF THE BIRDS OF THE CHATHAM ISLANDS

Species marked with † formerly bred on the Chatham Islands, but are no longer present (locally or globally extinct). The so-called Chatham Island sea eagle "*Haliaeetus australis*" is not included, as these bones are no longer considered to have come from the Chatham Islands (Holdaway *et al.* 2001). We also do not accept the undescribed teal, and fairy tern reported by Millener (1999) – see Holdaway *et al.* (2001). See Molloy *et al.* (2002) for explanation of the New Zealand Threat Classification System.

SPECIES NAME	STATUS NEW ZEALAND AND CHATHAM ISLANDS	NATIONAL THREAT RANK (CHATHAM BREEDERS ONLY)
Antipodean albatross *Diomedea [exulans] antipodensis*	New Zealand endemic, 2 pairs breeding Chathams	Range restricted
Southern royal albatross *D. epomophora*	New Zealand endemic, regular visitor Chathams	
Northern royal albatross *D. sanfordi*	Chatham Islands near endemic, breeding	Nationally vulnerable
Black-browed mollymawk *Thalassarche melanophrys*	New Zealand native, regular visitor	
Campbell Island mollymawk *T. impavida*	New Zealand endemic, regular visitor	
White-capped mollymawk *T. [cauta] steadi*	New Zealand endemic, 1 pair breeding Chathams	Range restricted
Salvin's mollymawk *T. salvini*	New Zealand native, a few breeding Chathams	Range restricted
Chatham Island mollymawk *T. eremita*	Chatham Islands endemic, breeding	Serious decline
Grey-headed mollymawk *T. chrysostoma*	New Zealand native, vagrant Chathams	

SPECIES NAME	STATUS NEW ZEALAND AND CHATHAM ISLANDS	NATIONAL THREAT RANK (CHATHAM BREEDERS ONLY)
Atlantic yellow-nosed mollymawk T. [chlororhynchos] chlororhynchos	Vagrant Chathams (only New Zealand records)	
Indian yellow-nosed mollymawk T. [chlororhynchos] carteri	New Zealand coloniser, 1 pair breeding Chathams	Coloniser
Buller's mollymawk T. bulleri	New Zealand endemic, vagrant Chathams	
Pacific mollymawk Thalassarche undescribed sp.	Chatham Islands near endemic, breeding	Range restricted
Light-mantled sooty albatross Phoebetria palpebrata	New Zealand native, vagrant Chathams	
Flesh-footed shearwater Puffinus carneipes	New Zealand native, vagrant Chathams	
Buller's shearwater P. bulleri	New Zealand endemic, regular visitor Chathams	
Sooty shearwater P. griseus	New Zealand native, breeding	Gradual decline
Short-tailed shearwater P. tenuirostris	Passage migrant New Zealand and Chathams	
Fluttering shearwater P. gavia	New Zealand endemic, vagrant Chathams	
Subantarctic little shearwater P. assimilis elegans	New Zealand native, breeding	Range restricted
Southern diving petrel Pelecanoides urinatrix chathamensis	New Zealand endemic subspecies, breeding	Not threatened
† South Georgian diving petrel P. georgicus	New Zealand native, locally extinct Chathams (subfossil)	Nationally critical
Grey petrel Procellaria cinerea	New Zealand native, regular visitor Chathams	

SPECIES NAME	STATUS NEW ZEALAND AND CHATHAM ISLANDS	NATIONAL THREAT RANK (CHATHAM BREEDERS ONLY)
Black petrel *P. parkinsoni*	New Zealand endemic, vagrant Chathams (subfossil)	
Westland petrel *P. westlandica*	New Zealand endemic, regular visitor Chathams	
White-chinned petrel *P. aequinoctialis steadi*	New Zealand native, regular visitor Chathams	
Kerguelen petrel *Lugensa brevirostris*	Regular visitor New Zealand, vagrant Chathams	
Cape pigeon *Daption capense capense*	Regular visitor New Zealand, vagrant Chathams	
Snares cape pigeon *D. c. australe*	New Zealand endemic subspecies, breeding	Range restricted
Antarctic petrel *Thalassoica antarctica*	Vagrant New Zealand and Chathams	
Antarctic fulmar *Fulmarus glacialoides*	Regular visitor New Zealand, vagrant Chathams	
Southern giant petrel *Macronectes giganteus*	Regular visitor New Zealand and Chathams	
Northern giant petrel *M. halli*	New Zealand native, breeding	Not threatened
Fairy prion *Pachyptila turtur*	New Zealand native, breeding	Not threatened
Chatham Island fulmar prion *P. crassirostris pyramidalis*	Chatham Islands endemic subspecies, breeding	Range restricted
Antarctic prion *P. desolata*	New Zealand native, vagrant Chathams	
Broad-billed prion *P. vittata*	New Zealand native, breeding	Not threatened
Blue petrel *Halobaena caerulea*	Regular visitor New Zealand, vagrant Chathams	
Black-winged petrel *Pterodroma nigripennis*	New Zealand native, breeding	Not threatened

SPECIES NAME	STATUS NEW ZEALAND AND CHATHAM ISLANDS	NATIONAL THREAT RANK (CHATHAM BREEDERS ONLY)
Chatham petrel *P. axillaris*	Chatham Islands endemic, breeding	Nationally endangered
Mottled petrel *P. inexpectata*	New Zealand endemic, vagrant Chathams	
Juan Fernandez petrel *P. externa*	Vagrant New Zealand, possible coloniser Chathams	
Kermadec petrel *P. neglecta*	New Zealand native, vagrant Chathams	
Grey-faced petrel *P. macroptera gouldi*	New Zealand endemic subspecies, vagrant Chathams	
Chatham Island taiko *P. magentae*	Chatham Islands endemic, breeding	Nationally critical
White-headed petrel *P. lessonii*	New Zealand native, vagrant Chathams	
Soft-plumaged petrel *P. mollis*	New Zealand native, vagrant Chathams	
† Gadfly petrel sp. *Pterodroma* undescribed sp.	Chatham Islands ?endemic, extinct (subfossil)	Extinct
Leach's storm petrel *Oceanodroma leucorhoa*	Vagrant New Zealand and Chathams	
Wilson's storm petrel *Oceanites oceanicus*	Passage migrant (rare New Zealand and Chathams)	
Grey-backed storm petrel *O. nereis*	New Zealand native, breeding	Not threatened
White-faced storm petrel *Pelagodroma marina maoriana*	New Zealand endemic subspecies, breeding	Not threatened
Black-bellied storm petrel *Fregetta tropica*	New Zealand native, vagrant Chathams	
King penguin *Aptenodytes patagonicus*	Vagrant New Zealand and Chathams (subfossil on Chathams)	
Yellow-eyed penguin *Megadyptes antipodes*	New Zealand endemic, vagrant Chathams	

SPECIES NAME	STATUS NEW ZEALAND AND CHATHAM ISLANDS	NATIONAL THREAT RANK (CHATHAM BREEDERS ONLY)
Chatham Island blue penguin *Eudyptula minor chathamensis*	Chatham Islands endemic subspecies, breeding	Range restricted
Eastern rockhopper penguin *Eudyptes chrysocome filholi*	New Zealand native, vagrant Chathams	
Moseley's rockhopper penguin *E. c. moseleyi*	Vagrant New Zealand and Chathams	
Snares crested penguin *E. robustus*	New Zealand endemic, vagrant Chathams	
Erect-crested penguin *E. sclateri*	New Zealand endemic, vagrant Chathams	
† Chatham Island crested penguin *Eudyptes* undescribed sp.	Chatham Islands endemic, extinct (subfossil)	Extinct
Australasian gannet *Morus serrator*	New Zealand native, vagrant Chathams	
Masked booby *Sula dactylatra*	New Zealand native, vagrant Chathams (subfossil)	
Black shag *Phalacrocorax carbo*	New Zealand native, breeding	Sparse
Little black shag *P. sulcirostris*	New Zealand native, vagrant Chathams	
Little shag *P. melanoleucos*	New Zealand native, vagrant Chathams	
Chatham Island shag *Leucocarbo onslowi*	Chatham Islands endemic, breeding	Range restricted
Pitt Island shag *Stictocarbo featherstoni*	Chatham Islands endemic, breeding	Range restricted
Lesser frigatebird *Fregata ariel*	Vagrant New Zealand and Chathams	
White-faced heron *Ardea novaehollandiae*	New Zealand native, breeding	Not threatened
Reef heron *E. sacra*	New Zealand native, vagrant Chathams	

SPECIES NAME	STATUS NEW ZEALAND AND CHATHAM ISLANDS	NATIONAL THREAT RANK (CHATHAM BREEDERS ONLY)
White heron *Casmerodius albus*	New Zealand native, vagrant Chathams	
Cattle egret *Bubulcus ibis*	Migrant to New Zealand, vagrant Chathams	
Australasian bittern *Botaurus poiciloptilus*	New Zealand native, vagrant Chathams	
† New Zealand little bittern *Ixobrychus novaezelandiae*	New Zealand endemic, extinct (subfossil)	Extinct
Glossy ibis *Plegadis falcinellus*	Vagrant New Zealand and Chathams	
Royal spoonbill *Platalea regia*	New Zealand native, vagrant Chathams	
Black swan *Cygnus atratus*	Re-introduced to New Zealand and Chathams	Re-introduced
Canada goose *Branta canadensis*	Introduced to New Zealand, vagrant Chathams	
Feral goose *Anser anser*	Introduced to New Zealand and Chathams, breeding	Introduced
Paradise shelduck *Tadorna variegata*	New Zealand endemic, vagrant Chathams	
Chestnut-breasted shelduck *T. tadornoides*	Vagrant New Zealand and Chathams	
† Chatham Island shelduck *Tadorna* undescribed sp.	Chatham Islands endemic, extinct (subfossil)	Extinct
Mallard *Anas platyrhynchos*	Introduced to New Zealand, breeding	Introduced
Grey duck *A. superciliosa*	New Zealand native, breeding	Serious decline
Grey teal *A. gracilis*	New Zealand native, vagrant Chathams	
† Brown teal *A. chlorotis*	New Zealand endemic, locally extinct (c.1915)	Nationally endangered

SPECIES NAME	STATUS NEW ZEALAND AND CHATHAM ISLANDS	NATIONAL THREAT RANK (CHATHAM BREEDERS ONLY)
† Australasian shoveler *A. rhynchotis*	New Zealand native, locally extinct (c.1925) and vagrant	Not threatened
† Chatham Island flightless duck *Pachyanas chathamica*	Chatham Islands endemic, extinct (subfossil)	Extinct
† New Zealand scaup *Aythya novaeseelandiae*	New Zealand endemic, locally extinct (subfossil)	Not threatened
† New Zealand merganser *Mergus australis*	New Zealand endemic, extinct (subfossil)	Extinct
† Scarlett's duck *Malacorhynchus scarletti*	New Zealand endemic, extinct (subfossil)	Extinct
Australasian harrier *Circus approximans*	New Zealand native, breeding	Not threatened
† New Zealand falcon *Falco novaeseelandiae*	New Zealand endemic, locally extinct (c.1890)	Nationally vulnerable
California quail *Callipepla californica*	Introduced to New Zealand and Chathams, breeding (rare)	Introduced
Buff weka *Gallirallus australis hectori*	New Zealand endemic introduced to Chathams	Not threatened
† Dieffenbach's rail *G. dieffenbachii*	Chatham Islands endemic, extinct (c.1840)	Extinct
† Chatham Island rail *Cabalus modestus*	Chatham Islands endemic, extinct (c.1900)	Extinct
† Hawkin's rail *Diaphorapteryx hawkinsi*	Chatham Islands endemic, extinct (subfossil)	Extinct
Spotless crake *Porzana tabuensis*	New Zealand native, ?breeding	Sparse
Marsh crake *P. pusilla*	New Zealand native, ?breeding	Sparse
Pukeko *Porphyrio melanotus*	New Zealand native, breeding	Not threatened
† Chatham Island coot *Fulica chathamensis*	Chatham Islands endemic, extinct (subfossil)	Extinct

SPECIES NAME	STATUS NEW ZEALAND AND CHATHAM ISLANDS	NATIONAL THREAT RANK (CHATHAM BREEDERS ONLY)
New Zealand pied oystercatcher *Haematopus finschi*	New Zealand endemic, vagrant Chathams	
Chatham Island oystercatcher *H. chathamensis*	Chatham Islands endemic, breeding	Nationally critical
Australasian pied stilt *Himantopus himantopus leucocephalus*	New Zealand native, breeding	Not threatened
Banded dotterel *Charadrius bicinctus bicinctus*	New Zealand endemic, breeding	Gradual decline
Mongolian plover *C. mongolus*	Vagrant New Zealand and Chathams	
Oriental plover *C. veredus*	Vagrant New Zealand and Chathams	
New Zealand shore plover *Thinornis novaeseelandiae*	New Zealand endemic, breeding	Nationally critical
Wrybill *Anarhynchus frontalis*	New Zealand endemic, vagrant Chathams	
Pacific golden plover *Pluvialis fulva*	Migrant New Zealand and Chathams	
Grey plover *P. squatarola*	Vagrant New Zealand and Chathams	
Spur-winged plover *Vanellus miles novaehollandiae*	New Zealand native, breeding	Not threatened
Turnstone *Arenaria interpres*	Migrant New Zealand and Chathams	
Chatham Island snipe *Coenocorypha pusilla*	Chatham Islands endemic, breeding	Range restricted
† Forbes' snipe *C. chathamica*	Chatham Islands endemic, extinct (subfossil)	Extinct
Lesser knot *Calidris canutus*	Migrant New Zealand and Chathams	

SPECIES NAME	STATUS NEW ZEALAND AND CHATHAM ISLANDS	NATIONAL THREAT RANK (CHATHAM BREEDERS ONLY)
Sanderling *C. alba*	Vagrant New Zealand and Chathams	
Curlew sandpiper *C. ferruginea*	Migrant to New Zealand, vagrant Chathams	
Sharp-tailed sandpiper *C. acuminata*	Migrant to New Zealand, vagrant Chathams	
Pectoral sandpiper *C. melanotos*	Vagrant New Zealand and Chathams	
Red-necked stint *C. ruficollis*	Migrant to New Zealand, vagrant Chathams	
Eastern curlew *Numenius madagascariensis*	Migrant to New Zealand, vagrant Chathams	
Asiatic whimbrel *N. phaeopus variegatus*	Migrant to New Zealand, vagrant Chathams	
American whimbrel *N. p. hudsonicus*	Vagrant New Zealand and Chathams	
Eastern bar-tailed godwit *Limosa lapponica baueri*	Migrant New Zealand and Chathams	
Black-tailed godwit *L. limosa*	Vagrant New Zealand and Chathams	
Wandering tattler *Tringa incana*	Vagrant New Zealand and Chathams	
Grey-tailed tattler *T. brevipes*	Vagrant New Zealand and Chathams	
Greenshank *T. nebularia*	Vagrant New Zealand and Chathams	
Marsh sandpiper *T. stagnatilis*	Vagrant New Zealand and Chathams	
Lesser yellowlegs *T. flavipes*	Vagrant New Zealand and Chathams	
Brown skua *Catharacta skua lonnbergi*	New Zealand native, breeding	Sparse

SPECIES NAME	STATUS NEW ZEALAND AND CHATHAM ISLANDS	NATIONAL THREAT RANK (CHATHAM BREEDERS ONLY)
South polar skua *C. maccormicki*	Passage migrant (rare New Zealand and Chathams)	
Arctic skua *Stercorarius parasiticus*	Migrant to New Zealand, vagrant Chathams	
Pomarine skua *S. pomarinus*	Vagrant New Zealand and Chathams	
Southern black-backed gull *Larus dominicanus*	New Zealand native, breeding	Not threatened
Red-billed gull *L. novaehollandiae scopulinus*	New Zealand endemic subspecies, breeding	Not threatened
Black-fronted tern *Sterna albostriata*	New Zealand endemic, vagrant Chathams (subfossil)	
Caspian tern *S. caspia*	New Zealand native, vagrant Chathams	
White-fronted tern *S. striata*	New Zealand native, breeding	Gradual decline
Antarctic tern *S. vittata*	New Zealand native, vagrant Chathams	
Little tern *S. albifrons*	Migrant to New Zealand, vagrant Chathams	
Arctic tern *S. paradisaea*	Passage migrant (rare New Zealand and Chathams)	
Parea *Hemiphaga chathamensis*	Chatham Islands endemic, breeding	Nationally critical
Rock pigeon *Columba livia*	Introduced to New Zealand, vagrant Chathams	
† Chatham Island kaka *Nestor* undescribed sp.	Chatham Islands endemic, extinct (subfossil)	Extinct
Chatham Island red-crowned parakeet *Cyanoramphus novaezelandiae chathamensis*	Chatham Islands endemic subspecies, breeding	Range restricted

SPECIES NAME	STATUS NEW ZEALAND AND CHATHAM ISLANDS	NATIONAL THREAT RANK (CHATHAM BREEDERS ONLY)
Forbes' parakeet *C. forbesi*	Chatham Islands endemic, breeding	Nationally endangered
Shining cuckoo *Chrysococcyx lucidus*	New Zealand native, breeding	Not threatened
Long-tailed cuckoo *Eudynamys taitensis*	New Zealand endemic, vagrant Chathams	
Fork-tailed swift *Apus pacificus*	Vagrant New Zealand and Chathams	
Sacred kingfisher *Todiramphus sancta*	New Zealand native, vagrant Chathams	
Skylark *Alauda arvensis*	Introduced to New Zealand and Chathams, breeding	Introduced
Welcome swallow *Hirundo tahitica neoxena*	New Zealand native, breeding	Not threatened
Australian tree martin *H. nigricans*	Vagrant New Zealand and Chathams	
Chatham Island pipit *Anthus novaeseelandiae chathamensis*	Chatham Islands endemic subspecies, breeding	Not threatened
Dunnock (hedge sparrow) *Prunella modularis*	Introduced to New Zealand, breeding	Introduced
Blackbird *Turdus merula*	Introduced to New Zealand and Chathams, breeding	Introduced
Song thrush *T. philomelos*	Introduced to New Zealand, breeding	Introduced
† Chatham Island fernbird *Bowdleria rufescens*	Chatham Islands endemic, extinct (c.1900)	Extinct
Chatham Island warbler *Gerygone albofrontata*	Chatham Islands endemic, breeding	Range restricted
Chatham Island fantail *Rhipidura fuliginosa penita*	Chatham Islands endemic subspecies, breeding	Not threatened

SPECIES NAME	STATUS NEW ZEALAND AND CHATHAM ISLANDS	NATIONAL THREAT RANK (CHATHAM BREEDERS ONLY)
Willie wagtail *R. leucophrys*	Vagrant Chathams (only New Zealand record)	
Chatham Island tomtit *Petroica macrocephala* *chathamensis*	Chatham Islands endemic subspecies, breeding	Nationally endangered
Black robin *P. traversi*	Chatham Islands endemic, breeding	Nationally critical
Silvereye *Zosterops lateralis*	New Zealand native, breeding	Not threatened
† Chatham Island bellbird *Anthornis melanocephala*	Chatham Islands endemic, extinct (c.1906)	Extinct
Chatham Island tui *Prosthemadera* *novaeseelandiae chathamensis*	Chatham Islands endemic subspecies, breeding	Nationally endangered
Yellowhammer *Emberiza citrinella*	Introduced to New Zealand, ?breeding	Introduced
Chaffinch *Fringilla coelebs*	Introduced to New Zealand, breeding	Introduced
Greenfinch *Carduelis chloris*	Introduced to New Zealand, breeding	Introduced
Goldfinch *C. carduelis*	Introduced to New Zealand, breeding	Introduced
Redpoll *C. flammea*	Introduced to New Zealand, breeding	Introduced
House sparrow *Passer domesticus*	Introduced to New Zealand, breeding	Introduced
Starling *Sturnus vulgaris*	Introduced to New Zealand and Chathams, breeding	Introduced
Rook *Corvus frugilegus*	Introduced to New Zealand, vagrant Chathams	
† Chatham Islands crow *C. moriorum*	Chatham Islands endemic, extinct (subfossil)	Extinct

Appendix 2

PLANT NAMES MENTIONED IN TEXT

COMMON NAME	SCIENTIFIC NAME
Bamboo rush	*Sporodanthus traversii*
Chatham Island akeake*	*Olearia traversii*
Chatham Island flax	*Phormium* aff. *tenax*
Clover	*Trifolium* spp.
Hoho	*Pseudopanax chathamicus*
Karaka	*Corynocarpus laevigatus*
Kopi	*Corynocarpus laevigatus*
Macrocarpa	*Cupressus macrocarpa*
Marram grass	*Ammophila arenaria*
Ngaio	*Myoporum laetum*
Pohuehue	*Muehlenbeckia australis*

*Unrelated to the akeake of mainland New Zealand (*Dodonaea* viscosa), which has been introduced to the Chatham Islands.*